Adaptation, Learning, and Optimization 14

Series Editors-in-Chief

Meng-Hiot Lim
Division of Circuits and Systems
School of Electrical & Electronic Engineering
Nanyang Technological University
Singapore 639798
E-mail: emhlim@ntu.edu.sg

Yew-Soon Ong
School of Computer Engineering
Nanyang Technological University
Block N4, 2b-39
Nanyang Avenue
Singapore 639798
E-mail: asysong@ntu.edu.sg

For further volumes:
http://www.springer.com/series/8335

Siddhartha Shakya and Roberto Santana (Eds.)

Markov Networks in Evolutionary Computation

Editors
Siddhartha Shakya
Business Modelling and Operational Transformation Practice
BT Innovate & Design
Ipswich
UK

Roberto Santana
Intelligent Systems Group
Faculty of Informatics
University of the Basque Country (UPV/EHU)
San Sebastian
Spain

ISSN 1867-4534
e-ISSN 1867-4542
ISBN 978-3-642-44494-4
ISBN 978-3-642-28900-2 (eBook)
DOI 10.1007/978-3-642-28900-2
Springer Heidelberg New York Dordrecht London

Softcover reprint of the hardcover 1st edition 2012

Printed on acid-free paper

Springer is part of Springer Science+Business Media (www.springer.com)

Preface

Siddhartha Shakya and Roberto Santana

1 Why Markov Networks in Evolutionary Computation?

In recent years, the field of heuristic optimisation has experienced an upsurge in the number of proposals that incorporate machine learning techniques to the search of optimal solutions. The true impact that the introduction of intelligent search strategies has in terms of search efficiency and problem knowledge discovery has been confirmed in several applications to research and industrial domains.

Estimation of distribution algorithms (EDAs) [3, 5, 6, 8] are a class of evolutionary algorithms (EAs) characterised by the use of probabilistic modelling and machine learning methods. Given their suitability to deal with complex optimisation problems and their scalability in many optimisation domains, EDAs have been proposed as a competent version of traditional genetic algorithms.

EDAs work by replacing the genetic operators used by other EAs by a probabilistic graphical model (PGM) [1, 2, 4, 7, 9] that serves as a statistical description of the best solutions visited during the search. Learning and sampling methods are respectively used to construct and sample the probabilistic model. By applying in each generation the steps of selection, learning, and sampling, EDAs are able to progressively move the focus of the search to promising areas of the search space.

A characteristic feature of EDAs is the type of probabilistic model they use. Since probabilistic modelling is very related to the representation of the solutions and to the characteristics of the optimisation problem, model selection is not always straightforward. A clever choice of the PGM may be critical in the solution of the

Siddhartha Shakya
Business Modelling and Operational Transformation Practice, BT Innovate & Design, Ipswich, UK
e-mail: sid.shakya@bt.com

Roberto Santana
Intelligent Systems Group, Faculty of Informatics, University of the Basque Country (UPV/EHU), San Sebastian, Spain
e-mail: roberto.santana@ehu.es

problem. Research on EDAs has mainly focused on the use of directed probabilistic models such as Bayesian networks. However, other classes of probabilistic models are a better or competitive alternative in many situations.

This book focuses on the analysis and application of Markov networks, and undirected models in general, as probabilistic models in EDAs. Markov networks are not merely a different alternative to directed PGMs. They support a number of attributes that make them particularly suitable in the context of evolutionary optimisation. In contrast to Bayesian networks and other directed models, no information about the direction of the interaction between the variables of the problems is required by the model. This feature is consistent with the characteristic situation of many optimisation problems for which the direction between two related variables is meaningless.

The book comprises a set of relevant contributions that explain the rationale behind the use of Markov networks in EDAs, analyse important issues for their design, and present a number of exemplary application of EDAs based on Markov networks. Readers interested in knowing about general and efficient optimisation approaches, based on the use of machine learning techniques, will find examples of how to conceive, analyse and enhance algorithms fitted to their problems. More experienced researchers from the evolutionary computation community, interested in getting a detailed account of the characteristics of these algorithms, will find in this book a panoply of hot research topics and current challenges in the area.

Markov networks have been profusely investigated in the field of statistical physics and there is a strong link between them and other application fields of machine learning techniques such as computer vision and sequence analysis. Researchers in these areas, which are not familiar with evolutionary algorithms, will find in this book an opportunity to discover an exciting application area for Markov network learning and sampling algorithms. Some of the chapters included in the book address the question of how to adapt currently available methods based on undirected models to deal with EDAs limitations. The work comprised in this book shows that research on EDAs could certainly inspire other developments in the study of Markov networks. Furthermore, the variety of available theoretical results in the field of Markov networks could provide a solid platform for better understanding of advanced evolutionary algorithms and the conception of more powerful EDAs based on undirected graphical models.

2 Book Organisation and Brief Overview of the Chapters Content

This book is organised in three main parts:

1. Introductory part.
2. Theoretical contribution or current research on Markov network EDAs.
3. Applications of EDAs based on Markov networks to real world problems.

The introductory part comprises five chapters authored by the editors of this book. Chapter 1 gives an introduction to PGMs, with emphasis on Markov networks and the algorithms used to learn and sample these models. Chapter 2 explains EDAs and review some of the most recent advances in research on these algorithms, devoting particular attention to EDAs based on Markov networks. The next three chapters in the introductory part describe in detail three paradigmatic EDAs based on Markov networks. Chapter 3 explains the Markov optimisation algorithm (MOA), a recently introduced EDA. MOA has a simple structure and incorporates advanced features from previous EDAs based on Markov networks. Chapter 4 describes the DEUM class of algorithms. A characteristic feature of these algorithms is that the fitness of the solutions is modelled as a Gibbs distribution defined on the model structure. In contrast, the MN-FDA and MN-EDA approaches discussed in chapter 5 where originally conceived to learn a statistical model of the solutions straight from the data. The five chapters included in this introductory part should provide the reader less familiar with EDAs with a basic understanding of the different ways for applying Markov networks in EDAs.

The second part of the book comprises a set of highly relevant and recent developments in the conception of EDAs based on Markov networks. This part is opened by chapter 6 where Mühlenbein analyses some of the most sensitive theoretical questions on the application of EDAs based on Markov networks to additively decomposable functions. Two of the questions considered are the derivation of convergence results for finite samples, and the relevance and complexity related to the use of the Gibbs sampling distribution. In chapter 7, Ponce-de-León and Díaz-Díaz introduce an algorithm that extends the representation capabilities of trees by means of adding edges based on a measure of edge relevance. The relationship between model learning and selection methods is also investigated. Brownlee et al. present in chapter 8 a deep analysis of the use of Markov networks as fitness function models. The chapter provides a detailed review of related approaches in evolutionary computation and illustrates with examples the potential of fitness modelling with Markov networks for the design of genetic operators and the extraction of a priori unknown problem structural information. In chapter 9, Mendiburu et al. investigate another hot topic in EDAs, the conception of alternative sampling methods that incorporate advanced inference techniques from the machine learning domain. The chapter illustrates how the implementation of loopy max-propagation on factor graphs constructed from the graphical models learnt by the EDA can improve the efficiency of these algorithms. In chapter 10, Karshenas et al. present the extension of EDAs based on Markov networks to continuous problems. The chapter shows the relationship between Gaussian distributions and Gaussian Markov random fields and introduces an algorithm for learning continuous marginal product models in the context of EDAs. This second part of the book closes with chapter 11 by Höens that illustrates the benefits from the creative application of concepts from statistical physics to evolutionary computation. The chapter focuses on improving the EDA probability estimation and sampling steps by means of subfunction merge strategies that cover a higher number of interactions, and the use of generalised belief propagation algorithms for efficient sampling.

The third and last part of the book comprises three chapters. In chapter 12, McCall et al review different applications of the DEUM approach. This chapter illustrates on the applicability of EDAs based on Markov networks to real world optimisation problems, covering domains such as chemotherapy optimisation, dynamic pricing, agricultural biocontrol, and case-based feature selection. Chapter 13, by Soto et al., addresses problems with a continuous representation. A molecular docking problem is solved by means of variants of copula-based EDAs. The modelling approach, a sophisticated PGM based on copulas is shown to outperform UMDA and other evolutionary algorithms for this challenging optimisation from bioinformatics. Chapter 14 by Handa investigates the use of EDAs based on conditional random fields for solving reinforcement learning problems. This chapter illustrates how the EDA application to this domain can resolve tasks that are generally difficult for traditional reinforcement learning algorithms.

References

1. Castillo, E., Gutierrez, J.M., Hadi, A.S.: Expert Systems and Probabilistic Network Models. Springer (1997)
2. Koller, D., Friedman, N.: Probabilistic Graphical Models: Principles and Techniques. The MIT Press (2009)
3. Larrañaga, P., Lozano, J.A. (eds.): Estimation of Distribution Algorithms. A New Tool for Evolutionary Computation. Kluwer Academic Publishers, Boston (2002)
4. Lauritzen, S.L.: Graphical Models. Oxford Clarendon Press (1996)
5. Lozano, J.A., Sagarna, R., Larrañaga, P.: Parallel Estimation of Distribution Algorithms. In: Larrañaga, P., Lozano, J.A. (eds.) Estimation of Distribution Algorithms. A New Tool for Evolutionary Computation, pp. 129–145. Kluwer Academic Publishers (2001)
6. Mühlenbein, H., Paaß, G.: From recombination of genes to the estimation of distributions I. Binary parameters. In: Ebeling, W., Rechenberg, I., Voigt, H.-M., Schwefel, H.-P. (eds.) PPSN 1996. LNCS, vol. 1141, pp. 178–187. Springer, Heidelberg (1996)
7. Pearl, J.: Probabilistic Reasoning in Intelligent Systems. Morgan Kaufman Publishers, Palo Alto (1988)
8. Pelikan, M., Goldberg, D.E., Lobo, F.: A survey of optimization by building and using probabilistic models. Computational Optimization and Applications 21(1), 5–20 (2002)
9. Whittaker, J.: Graphical Models in Applied Multivariate Statistics. Wiley Series in Probability and Mathematical Statistics, New York (1991)

Contents

List of Contributors

Adriel Álvarez
University of Havana. Cuba.
a.mosquera@lab.matcom.uh.cu

Concha Bielza
Technical University of Madrid. Spain.
mcbielza@fi.upm.es

Alexander Brownlee
Loughborough University. UK
A.E.I.Brownlee@lboro.ac.uk

Diana Carrera
University of Havana. Cuba.
d.carrera@lab.matcom.uh.cu

Elva Díaz-Díaz
University of Aguascalientes. Mexico
ediazd@correo.uaa.mx

Yasser González-Fernández
Institute of Cybernetics, Mathematics and Physics. Cuba.
ygf@icmf.inf.cu

Hisashi Handa
Okayama University. Japan.
handa@sdc.it.okayama-u.ac.jp

Robin Höns
AIS Institute. Germany.
robin@hoens.net

Hossein Karshenas
Technical University of Madrid. Spain.
hossein.karshenas@upm.es

Pedro Larrañaga
Technical University of Madrid. Spain.
pedro.larranaga@fi.upm.es

Jose A. Lozano
Univesity of the Basque Country (UPV/EHU). Spain.
ja.lozano@ehu.es

John McCall
Robert Gordon University. UK
j.mccall@rgu.ac.uk

Alexander Mendiburu (UPV/EHU)
Univesity of the Basque Country. Spain.
alexander.mendiburu@ehu.es

Yanely Milanés
University of Havana. Cuba
y.milanes@lab.matcom.uh.cu

Ernesto Moreno
Center of Molecular Immunology. Cuba.
emoreno@cim.sld.cu

Heinz Mühlenbein
AIS Institute. Germany.
heinz.muehlenbein@ais.fraunhofer.de

Alberto Ochoa
Institute of Cybernetics, Mathematics and Physics. Cuba.
ochoa@icmf.inf.cu

Gilbert Owusu
BT Innovate and Design. UK.
gilbert.owusu@bt.uk

Eunice Esther Ponce-de-Leon-Senti
University of Aguascalientes. Mexico.
eponce@correo.uaa.mx

Roberto Santana
University of the Basque Country (UPV/EHU). Spain.
roberto.santana@ehu.es

Siddhartha Shakya
Business Modelling and Operational Transformation Practice,
BT Innovate & Design, Ipswich, UK.
sid.shakya@bt.com

Marta Soto
Institute of Cybernetics, Mathematics and Physics. Cuba.
msoto@icmf.inf.cu

Part I
Introduction

Chapter 1
Probabilistic Graphical Models and Markov Networks

Roberto Santana and Siddhartha Shakya

Abstract. This chapter introduces probabilistic graphical models and explain their use for modelling probabilistic relationships between variables in the context of optimisation with EDAs. We focus on Markov networks models and review different algorithms used to learn and sample Markov networks. Other probabilistic graphical models are also reviewed and their differences with Markov networks are analysed.

1.1 Introduction

Probabilistic graphical models (PGMs) [5, 20, 24, 34, 44] have become common knowledge representation tools capable of efficiently representing and handling independence relationships. They are particularly suitable as a framework for dealing with problems in the presence or uncertainty. PGMs have been applied with excellent results to a variety of problem domains such as bioinformatics [23], communications [12], image analysis [25], etc.

A PGM usually encodes a representation of a joint generalised probability or density function corresponding to a set of random variables. This representation allows to considerably save the amount of memory needed to store the joint probability distribution by exploiting the conditional independence relations between the variables captured by the model. When knowledge is represented in terms of the probability dependencies contained in a graphical model, it can be used to infer the probability of an event given that some (possibly incomplete) information is available.

Roberto Santana
Intelligent Systems Group, Faculty of Informatics,
University of the Basque Country (UPV/EHU), San-Sebastian, Spain
e-mail: roberto.santana@ehu.es

Siddhartha Shakya
Business Modelling and Operational Transformation Practice,
BT Innovate & Design, Ipswich, UK
e-mail: sid.shakya@bt.com

S. Shakya and R. Santana (Eds.): Markov Networks in Evolutionary Computation, ALO 14, pp. 3–19.
springerlink.com

In the most general case a graphical model contains a quality component represented by a graph or graphical structure, and a quantitative part which specifies the numerical parameters used to represent the distribution. The graphical representation can be given in different forms (e.g. undirected and directed graphs, factor graphs, chain graphs, etc.) but in every case the semantics of the graph will capture certain dependence relationships between the variables. A PGM can be constructed by an expert with the aim to represent in the model some knowledge about a problem domain. However, this is in general a difficult task, particularly if the number of problem variables is very high or the definition of the relationships between the variables depends on subjective criteria. Automatic methods for learning PGMs from data are extensively used in these situations.

Once a PGM has been constructed, it can be used with different purposes. In most applications, PGMs are used to compute the marginal distribution of one or a few variables, a process that is known as inference. PGM applications in machine learning are mainly supported on the use of inference. The use of PGMs in optimisation, and particularly its application in the context of Markov network based EDAs exhibits a number of particular characteristics. In EDAs, PGMs are primarily used for the generation of new solutions (sampling) and the methods employed for learning the models are intensively applied, highlighting the relevance of the efficiency factor in this context of application.

In this chapter we provide an introduction to PGMs and modelling based on graphical models that is mainly oriented to the potential application of these models in the context of optimisation. This introduction has been conceived with the aim of helping the reader to understand the subsequent chapters of this book and therefore it is clearly biased in the treatment of the covered topics. Work on PGMs is vast and very diverse. Certain concepts and theoretical results are particularly suitable for one field and have less application to other domains. Our particular selection has been guided by three aims: 1) Emphasise the representational capabilities of PGMs and their adequacy to represent a variety of optimisation problems. 2) Review a set of concepts, theoretical results and algorithms that are required for understanding initial and current research on EDAs. 3) Highlight those issues related to the choice of PGMs, and the learning and sampling algorithms they use, that should be taken into consideration when addressing an optimisation problem using EDAs.

The other sections of this chapter are organised as follows. In the next section we introduce conditional independence graphs and explain how the graph structure is usually related with the underlying probability distribution it represents. This section also covers a number of relevant concepts from graph theory. Section 1.3 briefly explains the class of Bayesian networks. Section 1.4 introduces Markov networks and Gaussian Markov random fields and review the methods commonly used to learn and sample Markov networks. Section 1.7 explains other PGMs that have been applied in the context of EDAs.

1.2 Graphs and Independence Graphs

To give an intuition of the rationale behind the use of PGMs we present the case of general dependency graphs. In the next sections more formal definition of other classes of PGMs will be given. To explain how the qualitative component of PGMs is used to represent probabilistic relationships between the variables of a problem, we need to introduce the class of distributions and a number of concepts from graph theory.

Let X be a random variable. A value of X is denoted x. $\mathbf{X} = (X_1,\dots,X_n)$ will denote a vector of n random variables. We will use $\mathbf{x} = (x_1,\dots,x_n)$ to denote an assignment to the variables. S will denote a set of indexes in $N = \{1,\dots,n\}$, and $\mathbf{X}_S$ (respectively, $\mathbf{x}_S$) a subset of the variables of $\mathbf{X}$ (respectively, a subset of values of $\mathbf{x}$) determined by the indexes in S. The joint generalised probability distribution of $\mathbf{x}$ is represented as $\rho(\mathbf{X} = \mathbf{x})$ or $\rho(\mathbf{x})$. $\rho(\mathbf{x}_S)$ will denote the marginal generalised probability distribution for $\mathbf{X}_S$. We use $\rho(X_i = x_i \mid X_j = x_j)$ or, in a simplified form, $\rho(x_i \mid x_j)$, to denote the conditional generalised probability distribution of X_i given $X_j = x_j$.

If the set of random variables $\mathbf{X}$ is discrete, $\rho(\mathbf{X} = \mathbf{x}) = p(\mathbf{X} = \mathbf{x})$ or $p(\mathbf{x})$. $p(\mathbf{X} = \mathbf{x})$ is known as the joint probability mass function for $\mathbf{X}$. Similarly, $p(X_i = x_i)$ is the marginal mass probability function of X_i and $p(x_i \mid x_j)$ is the conditional mass probability of X_i given $X_j = x_j$.

If the set of random variables $\mathbf{X}$ is continuous, $\rho(\mathbf{X} = \mathbf{x}) = f(\mathbf{x})_{\mathbf{X}}$ or $f(\mathbf{x})$. $f(\mathbf{X} = \mathbf{x})$ is known as the joint density function of $\mathbf{X}$. Similarly, $f(x_i)_{X_i}$ is the marginal density function of X_i and $f(x_i \mid x_j)$ is the conditional density function of X_i given $X_j = x_j$.

A *graph* $G = (V, E)$ is defined by a non-empty finite set V of elements called vertices together with a possibly empty set E of pairs of vertices called edges. When edges are directed they are usually called arcs. Undirected graphs have only undirected edges. Similarly, directed graphs have only arcs.

Dependency graphs display the probability dependencies that exist between the random variables of a given probability distribution. Let P represent the dependence structure of the distribution. The relationship between P and its graphical representation G is given in terms of conditional independence (denoted with $\perp\!\!\!\perp_P$) and graphical separation (denoted with $\perp\!\!\!\perp_G$) [34].

A graph G is a dependency map (or D-map) of the probabilistic dependence structure of $\mathbf{X}$ if there is a one-to-one correspondence between the random variables in $\mathbf{X}$ and the nodes in V, such that for all disjoint subsets $\mathbf{X}_A$, $\mathbf{X}_B$, and $\mathbf{X}_C$ of $\mathbf{X}$ we have:

$$\mathbf{X}_A \perp\!\!\!\perp_P \mathbf{X}_B \mid \mathbf{X}_C \Longrightarrow \mathbf{X}_A \perp\!\!\!\perp_G \mathbf{X}_B \mid \mathbf{X}_C \tag{1.1}$$

Similarly, G is an independency map (or I-map) of P if

$$\mathbf{X}_A \perp\!\!\!\perp_P \mathbf{X}_B \mid \mathbf{X}_C \Longleftarrow \mathbf{X}_A \perp\!\!\!\perp_G \mathbf{X}_B \mid \mathbf{X}_C \tag{1.2}$$

G is said to be perfect map of P if it is both a D-map and an I-map, that is

$$\mathbf{X}_A \perp\!\!\!\perp_P \mathbf{X}_B \mid \mathbf{X}_C \Longleftrightarrow \mathbf{X}_A \perp\!\!\!\perp_G \mathbf{X}_B \mid \mathbf{X}_C \tag{1.3}$$

and in this case P is said to be isomorphic to G.

To emphasise the relationship between vertexes and variables, vertexes may be named after their variables: $V = \{X_1, \ldots, X_n\}$. When we refer to a variable, or a vertex or node, it will be clear from the context.

I-maps allows the definition of the graphical separation concept that implies that if two disjoint sets of nodes $\mathbf{X}_A$ and $\mathbf{X}_B$ are found to be separated by $\mathbf{X}_C$ in the graph they correspond to independent sets of variables conditioned on $\mathbf{X}_C$. The characterisation of separation varies for directed and undirected graphs [5, 34] and it plays an important role in the interpretability of graphical models.

1.2.1 Relevant Concepts from Graph Theory

There are a number of concepts from graph theory that are relevant in the analysis of PGMs. We discuss some of these concepts here.

Given a graph $G = (V, E)$, a *clique* in G is a fully connected subset of V. We reserve the letter C to refer to a clique. The collection of all cliques in G is denoted as $\mathscr{C}$. C is maximal when it is not contained in any other clique. C is the maximum clique of the graph if it is the clique in $\mathscr{C}$ with the highest number of vertexes.

A cycle in a graph is a vertex sequence $V_1, \ldots, V_m$ where $V_1 = V_m$ but all other pairs are distinct, and $\{V_i, V_{i+1}\} \in E$. A cycle is chordless if all pairs of vertexes that are not adjacent in the cycle are not neighbours. An undirected graph is said to be chordal if every cycle of length four or more has a chord.

A *junction graph* constructed from a set of maximal cliques is a graph where each node corresponds to a maximal clique, and there exists an edge between two nodes if their corresponding cliques overlap.

A fundamental property of chordal graphs is that their maximal cliques can be joined to form a tree, called the *junction tree*. A junction tree is an acyclic junction graph. In a junction tree any two cliques containing a node α are either adjacent in the junction tree, or connected by a chain made entirely of cliques that contain α. This property is called the *running intersection property* [24].

Figure 1.1a) shows an undirected graph with four maximal cliques and Figure 1.1b) shows an associated junction graph.

A key issue in the use of PGMs is the mapping between the topological characteristics of the dependency graph and the dependence relationships between the random variables. Some notions from graph theory has a parallel interpretation in terms of the dependencies. For instance, the neighbouring variables of X_i in an undirected dependency graph are sometimes referred to as *Markov Blanket* for X_i [32]. Similarly, some properties have alternative formulations. For example, given a dependency graph the following statements are equivalent [44]:

- $\mathbf{X}$ is decomposable.
- G is triangulated or chordal.
- A junction tree of G can be constructed.

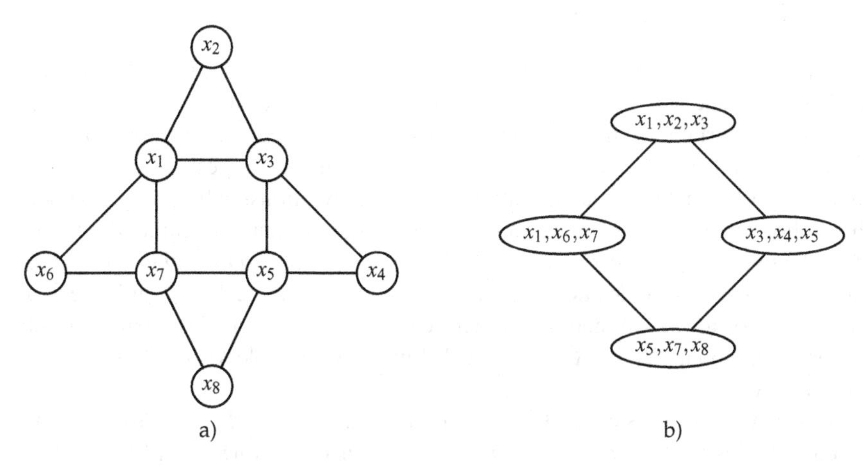

Fig. 1.1 a) Original undirected graph. b) Associated junction graph

1.3 Bayesian Networks

A Bayesian network is a pair (B, Θ), where B is the structure of the model and the Θ is a set of parameters of the model. The structure B is a *Directed Acyclic Graph (DAG)*[1], where each node corresponds to a variable in the modelled data set and each edge corresponds to a conditional dependency. A set of nodes Π_i is said to be the parent of X_i if there are edges from each variable in Π_i pointing to X_i. The parameter $\Theta = \{p(x_1|\Pi_1), p(x_2|\Pi_2), ..., p(x_n|\Pi_n)\}$ of the model is the set of conditional probabilities, where each $p(x_i|\Pi_i)$ is the set of probabilities associated with a variable $X_i = x_i$ given the different configuration of it's parent variables Π_i.

In general, given a set of variables $\mathbf{X} = \{X_1, X_2, .., X_n\}$ a joint probability distribution $p(\mathbf{X} = \mathbf{x})$ (or simply $p(\mathbf{x})$) for any Bayesian network is

$$p(\mathbf{x}) = \prod_{i=1}^{n} p(x_i|\Pi_i) \tag{1.4}$$

Bayesian network is a well established area of research within machine intelligence community, and also within wider artificial intelligence (AI) community. There is now a substantial literature in this field, with major publications including [24, 33, 34]. They have also been widely used in evolutionary computation area, and particularly in EDAs. Many well known EDAs use Bayesian networks to model the distribution. More on Bayesian networks and their application to EDA can be found in [22, 36].

---

[1] A DAG is a graph where each edge joining two nodes is a *directed edge*, and also there is *no cycle* in the graph i.e. it is not possible to start from a node and travelling towards the correct direction return back to the starting node.

1.4 Markov Networks

A Markov network is a pair (G, Ψ), where G is the structure and the Ψ is the parameter set of the network. G is an undirected graph where each node corresponds to a random variable in the modelled data set and each edge corresponds to conditional dependencies between variables. However, unlike Bayesian networks, the edges in Markov networks are undirected. Here, the relationship between two nodes should be seen as a *neighbourhood relationship*, rather than a parenthood relationship. We use $\mathcal{N} = \{N_1, N_2, ..., N_n\}$ to define a *neighbourhood system* on G, where each N_i is the set of nodes neighbouring to a node X_i. Figure 1.1a) can be seen as an example of a Markov network structure on 8 random variables. Here, variable X_1 has 4 neighbours $N_1 = \{X_2, X_3, X_6, X_7\}$. Similarly, variable X_2 has 2 neighbours, $N_2 = \{X_1, X_3\}$.

A Markov network is characterised in terms of a neighbourhood relationship between variables by its *local Markov property* known as *Markovianity* [3][26], which states that the conditional probability of a node X_i given the rest of the variables can be completely defined in terms of the conditional probability of the node given its Markov blanket. In terms of probability it can be written as

$$p(x_i|\mathbf{x} - \{x_i\}) = p(x_i|N_i) \tag{1.5}$$

A Markov network is also characterised in terms of *cliques* in the undirected graph by its global property, the joint probability distribution, and can be written as

$$p(\mathbf{x}) = \frac{1}{Z} \prod_{i=1}^{m} \psi_i(c_i) \tag{1.6}$$

Where, $\psi_i(c_i)$ (or more precisely $\psi_i(C_i = c_i)$) is a *potential function* on clique $C_i \in X$, and m is the number of cliques in the structure G. $Z = \sum_{x \in \Omega} \prod_{i=1}^{m} \psi_i(c_i)$ is the normalising constant known as the *partition function* which ensures that $\sum_{x \in \Omega} p(x) = 1$. Here, Ω is the set of all possible combination of the variables in X.

Equivalently, using Hammersley-Clifford theorem [18], the global Markov property can also be written in terms of Gibbs distribution as

$$p(\mathbf{x}) = \frac{e^{-U(\mathbf{x})/T}}{Z} \tag{1.7}$$

where,

$$Z = \sum_{(\mathbf{y}) \in \Omega} e^{-U(\mathbf{y})/T} \tag{1.8}$$

is a normalising constant, T is a parameter of the Gibbs distribution known as the *temperature* and $U(\mathbf{x})$ (or more precisely $U(\mathbf{X} = \mathbf{x})$) is known as the *energy* of the distribution.

Given an undirected graph, G, on $\mathbf{X}$, energy, $U(\mathbf{x})$, is defined as a sum of *potential functions* over the cliques, C_i, in G.

$$U(\mathbf{x}) = \sum_{i=1}^{m} u_i(c_i) \tag{1.9}$$

Here, $u_i(c_i)$ (or more precisely $u_i(C_i = c_i)$) is a potential function defined over a clique $C_i \in \mathbf{X}$. Equation (1.7), in terms of clique potential function, can also be written as

$$p(\mathbf{x}) = \frac{e^{-\sum_{i=1}^{m} u_i(c_i)/T}}{Z} \tag{1.10}$$

Note that the relationship between $\psi_i(c_i)$ in (1.6) and $u_i(c_i)$ in (1.10) is defined as

$$\psi_i(c_i) = e^{-u_i(c_i)/T} \tag{1.11}$$

The clique potential function $u_i(c_i)$, that captures the interaction between variables in the clique c_i, should be carefully defined in order to get a desired behaviour of the Markov network.

Usually Markov networks are also represented by

$$p(\mathbf{x}) = \frac{e^{-\sum_{i=1}^{m} w_i f_i(\mathbf{x})/T}}{Z} \tag{1.12}$$

where a feature $f_i(\mathbf{x})$ may be any real-valued function of the state and parameters w_i are called weights.

1.4.1 Markov Networks of Bounded Complexity

Graphical models can be classified into decomposable and non decomposable models. The class of decomposable models can be represented by means of both directed and undirected graphs. Decomposable models are important because exact factorisations of the distribution can be directly derived from them. In particular, decomposable models or bounded complexity are regularly used in EDAs since they allow feasible approximations of the joint probability distributions.

The complexity of a graphical model is usually associated to the dimension of the structural components and the number of parameters needed to store the factorised distribution. The model complexity influences the expressivity of the probabilistic model to represent the relationships between the variables and also the computational cost of the algorithms used to learn the model and make inference.

According to the complexity of the models, PGMs can be divided into three classes: Univariate models, bivariate models and multivariate models according to whether they do not represent any type of conditional dependencies between the variables, represent only pairwise dependencies or are able to represent multivariate dependencies, as is the case of Markov networks and Bayesian networks.

In Figure 1.2a) the structure of a Markov network corresponding to a probability $p(x_1, \cdots, x_7)$ is shown. The undirected graphs shown in Figure 1.2b-d) are all chordal and correspond to different approximate factorisations of the PGM shown in Figure 1.2a). In the univariate approximation (Figure 1.2b)) we assume all the variables are independent. The corresponding joint probability is $p(\mathbf{x}) = \prod_{i=1}^{7} p(x_i)$. A model that considers a higher number of dependencies is the tree model (Figure 1.2c)), where each variable can depend at most on another variable, and the graph can not have cycles. In this case the corresponding value of the joint probability associated to a vector $\mathbf{x}$ is $p(\mathbf{x}) = P(x_4) \cdot \prod_{\substack{i=1 \\ i \neq 4}}^{7} p(x_i|x_4)$. Finally, the example presented in Figure 1.2d) shows the case of a chordal subgraph of the Markov network structure shown in Figure 1.2a). This chordal subgraph can be represented using a junction tree. The approximate factorised joint probability is $p(x) = P(x_1, x_2, x_3, x_4) \cdot P(x_5|x_2, x_4) \cdot P(x_6, x_7|x_4, x_5)$.

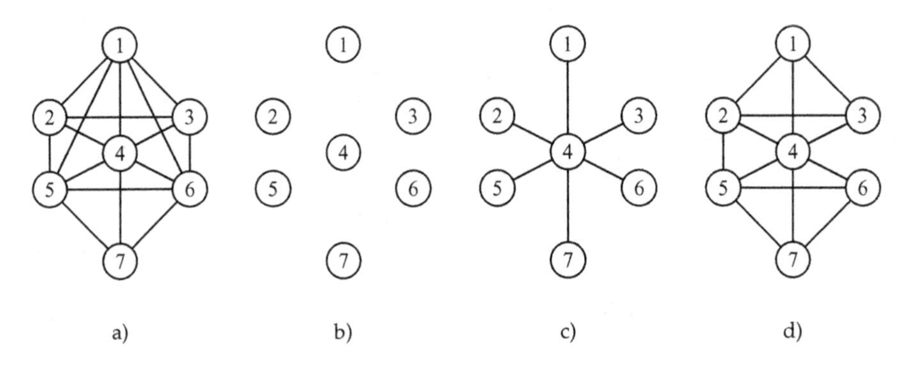

Fig. 1.2 a) Markov network. b) Univariate Model. c) Tree shaped network. d) Chordal subgraph.

1.4.2 Gaussian Markov Random Fields

A joint Gaussian distribution $\mathscr{N}(\mu, \Sigma)$ for n random variables is defined with two parameters: μ is the n-dimensional vector of mean values for each variable, and Σ is a positive-definite and symmetric $n \times n$ covariance matrix.

$$\mu = E(\mathbf{X})$$
$$\Sigma = E((\mathbf{X} - \mu)^{\mathbf{T}}(\mathbf{X} - \mu)) = \mathbf{E}(\mathbf{X}^{\mathbf{T}}\mathbf{X}) - \mu^{\mathbf{T}}\mu$$

The information form representation of a joint Gaussian distribution $\mathscr{N}(\mu, \Sigma)$, also known as the canonical or natural form is defined as

$$p(\mathbf{x}) = \mathscr{N}^{-1}(\mathbf{h}, \Theta) = \frac{e^{-\frac{1}{2}\mathbf{h}\Theta^{-1}\mathbf{h}^T}}{\sqrt{(2\pi)^n |\Theta^{-1}|}} e^{-\frac{1}{2}\mathbf{x}\Theta\mathbf{x}^T + \mathbf{x}\mathbf{h}^T} \tag{2}$$

where $\mathbf{h} = \mu\Sigma^{-1}$ is called the potential vector and $\Theta = \Sigma^{-1}$ is the inverse covariance matrix, known as the *precision*, concentration or information matrix. In this representation of the joint Gaussian distribution the precision matrix is positive definite.

A Gaussian Random Markov Field (GMRF) [38] is a normal distributed random vector that satisfies a number of Markov properties. More formally, a random variable $\mathbf{X} = (X_1, \ldots, X_n)^\top \in \mathcal{N}(\mu, \Sigma)$ is called a GMRF if the joint distribution for $\mathbf{x}$ satisfies $p(x_i \mid \mathbf{x} - \{x_i\}) = p(x_i \mid x_{N_i}) \forall i$.

In a GMRF, the following properties are equivalent:

1. X_i and X_j are conditionally independent
2. The corresponding element in the precision matrix, $\Theta_{i,j}$ is zero.
3. X_i and X_j are not neighbours.

1.5 Learning Markov Networks

Learning Markov networks of discrete random variables is a computationally intractable problem. The reason is the exponential complexity of computing the partition function. Nevertheless, several algorithms have been proposed to learn MNs from data. In this section, we describe some of these algorithms and discuss some of their characteristics and shortcomings. We focus on algorithms that have been proposed in domains different to optimisation. Learning algorithms used by Markov network based EDAs are covered in chapters 3, 4, and 5 of this book.

There are currently three main approaches to learn Markov models from data: Score+search methods, algorithms based on independence tests (also called constrained methods) and regularisation-based methods. A completely different alternative strategy is to do learning in the reduced space of decomposable Markov models. In this chapter we review Markov network learning algorithms from all these classes.

1.5.1 Learning of Discrete Markov Networks

1.5.1.1 Score+Search Methods

Score-based approaches associate a metric to each network. The metric will depend on the complexity of the graph and the goodness of fit of the graph to the data. Search procedures are employed to generate candidate models with the aim of finding a Markov network that maximises the score. The approach is usually applied in the context of methods that learn characterisations of Markov networks in terms of feature sets [10] (see Equation (1.12)). The problem of feature induction for Markov networks is similar to the problem of rule induction for classification. Among the learning algorithms based on score+search methods are:

- Top-down algorithm for inducing features of random fields [10]: Incrementally constructs an increasingly detailed field to approximate a reference distribution. A greedy algorithm determines how features are incrementally added to the field and an iterative scaling algorithm is used to estimate the optimal values of the weights.
- Feature induction algorithm for linear chain conditional random fields [28]: The method iteratively constructs feature conjunctions that would significantly increase conditional log-likelihood if added to the model. To adjust the parameters

of the Markov network a quasi-Newton method is used. This algorithm can also be classified as a top-down approach.

- Bottom-up learning of Markov network structure (BLM) [9]: The algorithm starts with each complete training example as a long feature and repeatedly generalises a feature to match its k nearest examples by dropping variables. In this way testing many feature variations with no support in the data is avoided.

1.5.1.2 Methods Based on Independence Tests

Algorithms based on independence tests work by conducting a series of statistical conditional independence tests toward the goal of restricting the number of possible structures to one, thus inferring that structure as the only possibly correct one [14]. In general, these algorithms work by independently computing the Markov blanket of each variable. The Markov network of any strictly positive distribution can be constructed by connecting each variable to all members of its Markov blanket [35]. A recent survey [40] covers Markov network learning algorithms based on independence tests. Among the representative algorithms based on independence tests are:

- The Grow-Shrink Markov Inference-based Markov network algorithm (GSIMN) [4]: It uses the axioms that govern the probabilistic independence relationships to reduce the number of unnecessary independence tests. It improves the efficiency over simple application of statistical tests. However it uses a rigid and heuristic ordering of the execution of the tests.
- Dynamic grow shrink inference-based Markov network algorithm (DGSIMN) [14]: This algorithm uses the same approach that GSIMN but uses a principled strategy, dynamically selecting the locally optimal test. As a result, it achieves an overall decrease in the computational requirement of the algorithm.
- The particle filter Markov network algorithm (PFMN) [27]: It uses a particle filter (sequential Monte Carlo) method to maintain a population of Markov network structures that represent the posterior probability distribution over structures, given the outcomes of the tests performed. By selecting the maximally informative test, the cost of the statistical tests conducted on data is decreased.

1.5.1.3 Regularisation Methods

Regularisation-based methods simultaneously learn the structure and the parameters of the model. The penalisation component acts as a regularisation parameter that helps to identify the structure of the model.

- Regularised model learning of pair-wise MRFs [37]: The neighbourhood of any given node is estimated by performing logistic regression subject to an $l1$ -constraint. Under certain conditions it is proved that consistent neighbourhood selection can be obtained with exponentially decaying error.

- Joint structure estimation of the MN parameters [17]: The likelihood function is estimated by a pseudolikelihood function and a joint l_1-penalised logistic regression method that simultaneously estimates the parameters for each of the n variables.

1.5.1.4 Learning of Decomposable Markov Networks

An alternative way of learning Markov networks is by constructing approximations of bounded complexity as those described in Section 1.4.1. The number of parameters needed to represent the class of decomposable Markov networks of small complexity is smaller and this class of Markov networks are also easier to sample. Among the algorithms that learn decomposable MNs are:

- Thin junction tree learning algorithm [2]: It learns models that are characterised by an upper bound on the size of the maximal cliques (no more than two nodes). Greedy search is used to find the set of features that enables the best possible fit to the distribution under the constraint on the clique sizes.
- Polynomial algorithm for PAC-learning of bounded-treewidth junction trees [6]: It learns limited-treewidth junction trees with strong intra-clique dependencies. The algorithm is based on the computation of the mutual information and is guaranteed to find a junction tree that is close in KL divergence to the true distribution.
- High-treewidth Markov network algorithm [16]: This feature-based learning algorithm allows to learn a much broader class of decomposable Markov networks than thin-treewidth algorithms. By exploiting context-specific independence and determinism in the domain, the algorithm allows to obtain Markov networks with larger treewidth.

There are several work on EDAs that learn restricted classes of decomposable Markov networks from data. We postpone the analysis of these algorithms to chapters 2 and 5.

1.5.2 Learning of Gaussian Markov Random Fields

The problem of learning GMRFs is easier than the Markov network learning for discrete variables. For GMRFs, the graph structure can be recovered by estimating the corresponding inverse covariance matrix. Another possibility is to learn the network directly without first estimating the inverse covariance matrix.

Among the methods use to learn Gaussian Markov random fields are:

- Sparse INverse COvariance selection problem (SINCO) [39]: It uses coordinate ascent, in a greedy manner, to optimise one diagonal or two symmetric off-diagonal elements of the precision matrix at each step. The method preserves the sparsity of the network structure and avoids introducing unnecessary (small) nonzero elements.

- Sparse covariance selection [8]: The problem of finding the edges of the GMRF is addressed by discovering the patterns of zeroes in the inverse covariance matrix. An estimation of a convex relaxation of the likelihood function is achieved using two different optimisation approaches. The methods seeks to trade-off the log-likelihood of the solution with the number of zeroes in its inverse.
- Learning sparse GMRFs [11]: The algorithm is able to find GMRFs in which certain groups of edges should all be penalised together. l_1-norm is used as a way to regularise the inverse covariance matrix.
- Neighbourhood selection with the Lasso penalty [30] : The algorithm relies on optimisation of a convex function, applied consecutively to each node in the graph. Neighbourhood selection estimates the conditional independence restrictions separately for each node in the graph and is hence equivalent to variable selection for Gaussian linear models. The neighbourhood selection scheme is consistent for sparse high-dimensional graphs.

1.6 Sampling Markov Networks

Sampling is used as a method for approximate inference in traditional applications of PGMs. The objective in this context is to obtain an approximation of the probability $p(\mathbf{x})$ to compute expectations from the probability. In EDAs, the goal of using sampling is not to compute a particular expectation of the distribution but to actually generate new solutions that could improve the current best found solutions. Sampling from $p(\mathbf{x})$ guarantees to exploit the information represented in the distribution about the good solutions. It should also allow to explore new areas of the search space that have not been exploited yet.

Sampling methods used for MNs can be roughly divided into two groups of algorithms:

1. Forward samplers: Given a topological order of nodes in the MN, each variable is sampled according to its conditional distribution given its ancestors in the ordering.
2. Markov blanket samplers: Each variable is sampled given assignments to its Markov blanket.

Forward samplers comprise forward sampling [41], rejection sampling [43], and likelihood weighting [13]. They require an ordering of the variables and therefore are mainly applied to tree Markov networks. Markov blanket samplers can be applied to any Markov network. Gibbs sampling [15] belongs to this class and is the sampling method that has been most frequently applied to EDAs based on Markov networks.

The main limitation of Gibbs sampling and other Markov blanket samplers is that it is computationally costly. Many sampling steps are required to obtain accurate estimates from the target probability distribution $p(\mathbf{x})$. More details on the application of Gibbs sampling to EDAs based on Markov networks are given in chapter 3.

1.7 Other Probabilistic Graphical Models

Directed and undirected graphs are just two examples of the variety of graphical representations and semantics available in field of PGMs. Some other models like chain graphs [7] can combine directed and undirected edges. In this section we cover three classes of more sophisticated graphical models that have application in EDAs. Mixtures of distributions, factor graphs and region graphs are reviewed. Examples of EDAs that use these PGMs are respectively discussed in chapters 5, 6 and 11.

1.7.1 Mixture Models

Probabilistic modelling by finite mixture of distributions [29] concerns modelling a statistical distribution by a mixture (or weighted sum) of other distributions. A mixture $q^m(\mathbf{x})$ of distributions $p_j(\mathbf{x})$ is defined to be a distribution of the form:

$$q^m(\mathbf{x}) = \sum_{j=1}^{m} \lambda_j p_j(\mathbf{x}) \tag{1.13}$$

with $\lambda_j > 0$, $j = 1,\ldots,m$, $\sum_{j=1}^{m} \lambda_j = 1$.

The $p_j(\mathbf{x})$ are called mixture components, and the λ_j are called mixture coefficients. m is the number of components of the mixture.

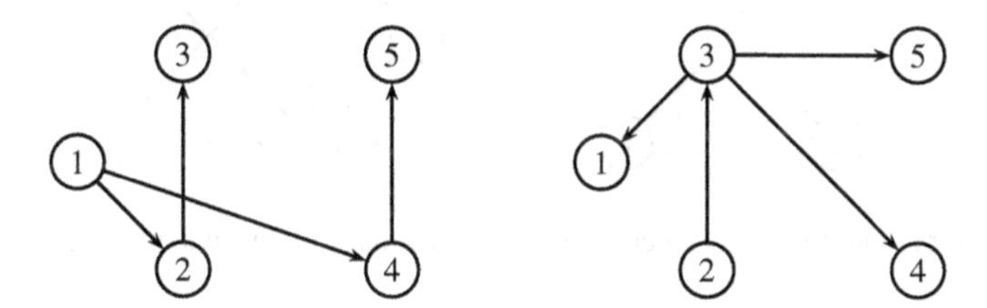

Fig. 1.3 The structure of a mixture of trees with 5 variables and 2 components.

The mixture components may have different structures. In Figure 1.3, the structure of a mixture of trees with 5 variables and 2 components is shown.

A mixture of distributions can be viewed as containing an unobserved choice variable $\mathbf{Z}$ which takes value $j \in \{1,\ldots,m\}$ with probability $p(\mathbf{z} = j) = \lambda_j$. In some cases the choice variable Z is known.

Probabilistic modelling based on mixtures of distributions has been used in many domains. Two of the most frequent applications are data clustering and approximation of probability distributions [29]. Mixtures are specially suited for modelling problems that exhibit complex interactions between their variables.

1.7.2 Factor Graphs

A *factor graph* [21] is a bipartite graph that can serve to represent the factorised structure of a distribution. It has two types of nodes: variable nodes (represented as a circle), and factor nodes (represented as a square). In the graphs, factor nodes are named by capital letters starting from A, and variable nodes by numbers starting from 1. Variable nodes are indexed with letters starting with i, and factor nodes with letters starting with a.

The existence of an edge connecting variable node i to factor node a means that x_i is an argument of function f_a in the referred factorisation. Figure 1.4 (left) shows a factor graph with two factor nodes and five variable nodes. The associated undirected graph (right) has two maximal cliques.

In [1], Gibbs distributions are associated with factor graphs. A factor f with scope $\mathbf{X}_S$ is a mapping from $\mathbf{x}_S$ to $\mathbb{R}^+$. A Gibbs distribution $p(\mathbf{x})$ is associated with a set of factors $\{f_a\}_{a=1}^m$ with scopes $\{\mathbf{X}_{S_a}\}_{a=1}^m$, such that

$$p_f(\mathbf{x}) = \frac{1}{Z}\prod_{a=1}^{m} f_a(\mathbf{x}_{S_a}) \tag{1.14}$$

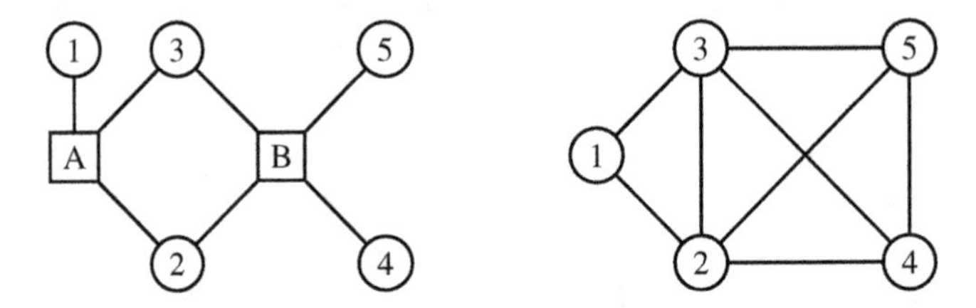

Fig. 1.4 Factor graph (left) and associated undirected graph with two maximal cliques (right).

1.7.3 Region Graphs

Yedidia et al. [45] introduced the concept of region graphs to define more precise free energy approximations of statistical systems. Usually, regions are defined to comprise the sets of elements that form local loops or where local correlations between the regions elements exist.

More formally, a region graph $\mathscr{R}$ is defined over a factor graph. It is formed by regions $\{\alpha, \gamma, \mu, \cdots\}$ and the edges $\{(\alpha, \gamma), (\mu, \nu), \cdots\}$ between regions [45]. A region α contains some variable nodes and function nodes of the factor graph, it satisfies the condition that, if a function node $a \in \alpha$, then all the variable nodes connected to a are in α.

For the same factor-graph it is usually possible to construct many different region graphs which all satisfy the region-graph condition. The factor-graph itself can be turned into the simplest region-graph with two types of regions (each small region is formed by a single variable node, and each large region is formed by a function node and all the connected variable nodes).

A graph region based decomposition is an asset of regions and an associated set of *counting numbers* U which is formed by one counting number c_R for each $R \in \mathscr{R}$. c_R will always be an integer, but might be zero or negative for some R.

To be valid, a decomposition must satisfy a number of constraints related to the regions and the counting numbers. There are several ways to construct valid region-based approximations. One of these approaches is the clustering variation method (CVM) introduced in [19, 31].

In CVM, $\mathscr{R}$ is formed by an initial set of regions $\mathscr{R}_0$ such that all the nodes are in at least one region of $\mathscr{R}_0$, and any other region in $\mathscr{R}$ is the intersection of one or more of the regions in $\mathscr{R}$. The set of regions $\mathscr{R}$ is closed under intersection, and can be ordered as a poset.

We say that a set of regions $\mathscr{R}$, and counting numbers c_R give a valid region based graph decomposition when for every variable X_i:

$$\sum_{\substack{R \in \mathscr{R} \\ X_i \subseteq X_R}} c_R = 1 \tag{1.15}$$

In chapters 5 and 11 applications of region graphs in the context of EDAs are discussed.

Acknowledgements. This work has been partially supported by the Saiotek and Research Groups 2007-2012 (IT-242-07) programs (Basque Government), TIN2010-14931 and Consolider Ingenio 2010 - CSD 2007 - 00018 projects (Spanish Ministry of Science and Innovation).

References

1. Abbeel, P., Koller, D., Ng, A.Y.: Learning factor graphs in polynomial time and sample complexity. Journal of Machine Learning Research 7, 1743–1788 (2006)
2. Bach, F.R., Jordan, M.I.: Thin junction trees. In: Proceedings of the Conference Advances in Neural Information Processing Systems 20, NIPS, pp. 569–576. MIT Press (2002)
3. Besag, J.: Spatial interactions and the statistical analysis of lattice systems (with discussions). Journal of the Royal Statistical Society 36, 192–236 (1974)
4. Bromberg, F., Margaritis, D., Honavar, V.: Efficient Markov network structure discovery using independence tests. Journal of Artificial Intelligence Research 35, 449–484 (2009)
5. Castillo, E., Gutierrez, J.M., Hadi, A.S.: Expert Systems and Probabilistic Network Models. Springer (1997)
6. Chechetka, A., Guestrin, C.: Efficient principled learning of thin junction trees. In: Proceedings of the Conference Advances in Neural Information Processing Systems 20, NIPS. MIT Press (2008)
7. Cox, D., Wermuth, N.: Linear dependencies represented by chain graphs. Statistical Science 8(3), 204–218 (1993)
8. d'Aspremont, A., Banerjee, O., Ghaoui, L.: First-order methods for sparse covariance selection. SIAM Journal on Matrix Analysis and Applications 30(1), 56–66 (2008)

9. Davis, J., Domingos, P.: Bottom-up learning of Markov network structure. In: Proceedings of the Twenty-Seventh International Conference on Machine Learning. ACM Press (2010)
10. Della Pietra, S., Della Pietra, V., Lafferty, J.: Inducing features of random fields. IEEE Transactions on Pattern Analysis and Machine Intelligence 19(4), 380–393 (1997)
11. Duchi, J., Gould, S., Koller, D.: Projected subgradient methods for learning sparse Gaussians. In: Proceedings of the 24th Annual Conference on Uncertainty in Artificial Intelligence, UAI 2008 (2008)
12. Frey, B.: Graphical Models for Machine Learning and Digital Communication. MIT Press (1998)
13. Fung, R., Chang, K.: Weighting and integrating evidence for stochastic simulation in Bayesian networks. Uncertainty in Artificial Intelligence 5, 209–219 (1989)
14. Gandhi, P., Bromberg, F., Margaritis, D.: Learning Markov network structure using few independence tests. In: Proceedings of the SIAM Conference on Data Mining, pp. 680–691 (2008)
15. Geman, S., Geman, D.: Stochastic relaxation, Gibbs distributions and the Bayesian restoration of images. In: Fischler, M.A., Firschein, O. (eds.) Readings in Computer Vision: Issues, Problems, Principles, and Paradigms, pp. 564–584. Kaufmann, Los Altos (1987)
16. Gogate, V., Austin, W., Domingos, W.: Learning efficient Markov networks. In: Proceedings of the Conference on Neural Information Processing Systems (NIPS 2010). MIT Press (2010)
17. Guo, J., Levina, E., Michailidis, G., Zhu, J.: Joint structure estimation for categorical Markov networks (2010) (summited), `http://www.stat.lsa.umich.edu/~elevina`
18. Hammersley, J.M., Clifford, P.: Markov fields on finite graphs and lattices (1971) (unpublished)
19. Kikuchi, R.: A theory of cooperative phenomena. Physical Review 81(6), 988–1003 (1951)
20. Koller, D., Friedman, N.: Probabilistic Graphical Models: Principles and Techniques. MIT Press (2009)
21. Kschischang, F.R., Frey, B.J., Loeliger, H.A.: Factor graphs and the sum-product algorithm. IEEE Transactions on Information Theory 47(2), 498–519 (2001)
22. Larrañaga, P.: An Introduction to probabilistic graphical models. In: Estimation of Distribution Algorithms. A New Tool for Evolutionary Computation, pp. 25–54. Kluwer Academic Publishers, Boston (2002)
23. Larrañaga, P., Calvo, B., Santana, R., Bielza, C., Galdiano, J., Inza, I., Lozano, J.A., Armañanzas, R., Santafé, G., Pérez, A., Robles, V.: Machine learning in bioinformatics. Briefings in Bioinformatics 7, 86–112 (2006)
24. Lauritzen, S.L.: Graphical Models. Oxford Clarendon Press (1996)
25. Li, S.: Markov Random Field Modeling in Image Analysis. Springer-Verlag New York Inc. (2009)
26. Li, S.Z.: Markov Random Field modeling in computer vision. Springer (1995)
27. Margaritis, D., Bromberg, F.: Efficient Markov network discovery using particle filters. Computational Intelligence 25(4), 367–394 (2009)
28. McCallum, A.: Efficiently inducing features of conditional random fields. In: Proceedings of the Nineteenth Conference on Uncertainty in Artificial Intelligence (UAI 2003), pp. 403–410 (2003)
29. McLachlan, G., Peel, D.: Finite Mixture Models. John Wiley & Sons (2000)

30. Meinshausen, N., Bühlmann, P.: High-dimensional graphs and variable selection with the lasso. Annals of Statistics 34, 1436–1462 (2006)
31. Morita, T.: Formal structure of the cluster variation method. Progress of Theoretical Physics Supplements 115, 27–39 (1994)
32. Murphy, K.: Dynamic Bayesian Networks: Representation, Inference and Learning. PhD thesis, University of California, Berkeley (2002)
33. Neapolitan, R.E.: Learning Bayesian Networks. Prentice-Hall, Upper Saddle River (2003)
34. Pearl, J.: Probabilistic Reasoning in Intelligent Systems. Morgan Kaufman Publishers, Palo Alto (1988)
35. Pearl, J., Paz, A.: Graphoids: A graph-based logic for reasoning about relevance relations. Technical Report R–53–L, Cognitive Systems Laboratory, University of California, Los Angeles (1985)
36. Pelikan, M.: Hierarchical Bayesian Optimization Algorithm. Toward a New Generation of Evolutionary Algorithms. STUDFUZZ, vol. 170. Springer, Heidelberg (2005)
37. Ravikumar, P., Wainwright, M., Lafferty, J.: High-dimensional ising model selection using l1-regularized logistic regression. The Annals of Statistics 38(3), 1287–1319 (2010)
38. Rue, H., Held, L.: Gaussian Markov Random Fields: Theory and Applications, vol. 104. Chapman & Hall (2005)
39. Scheinberg, K., Rish, I.: Learning sparse Gaussian Markov networks using a greedy coordinate ascent approach. In: Machine Learning and Knowledge Discovery in Databases, pp. 196–212 (2010)
40. Schlüter, F., Bromberg, F.: A survey on independence-based Markov networks learning, arXiv.org, arXiv:1108.2283 (2011)
41. Shachter, R., Peot, M.: Simulation approaches to general probabilistic inference on belief networks. In: Proceedings of the Fifth Annual Conference on Uncertainty in Artificial Intelligence, pp. 221–234. North-Holland Publishing Co., Amsterdam (1990)
42. Shakya, S.: DEUM: A Framework for an Estimation of Distribution Algorithm based on Markov Random Fields. PhD thesis, The Robert Gordon University, Aberdeen, UK (April 2006)
43. Von Neumann, J.: Various techniques used in connection with random digits. Applied Math Series 12(36-38), 1 (1951)
44. Whittaker, J.: Graphical Models in Applied Multivariate Statistics. Wiley Series in Probability and Mathematical Statistics, New York (1991)
45. Yedidia, J.S., Freeman, W.T., Weiss, Y.: Constructing free energy approximations and generalized belief propagation algorithms. Technical Report TR-2002-35, Mitsubishi Electric Research Laboratories (August 2002)

Chapter 2
A Review of Estimation of Distribution Algorithms and Markov Networks

Siddhartha Shakya and Roberto Santana

Abstract. This chapter reviews some of the popular EDAs based on Markov Networks. It starts by giving introduction to general EDAs and describes the motivation behind their emergence. It then categorises EDAs according to the type of probabilistic models they use (directed model based, undirected model based and common model based) and briefly lists some of the popular EDAs in each categories. It then further focuses on undirected model based EDAs, describes their general workflow and the history, and briefly reviews some of the popular EDAs based on undirected models. It also outlines some of the current research work in this area.

2.1 Introduction

The evolution process in Evolutionary Algorithms (EA) can be seen as a combination of two processes: Selection and Variation. Selective pressure favours the evolution of high-quality solutions. Variation helps to explore the search space and exploits those regions containing better solutions. An important factor in the success of this process is the linkage between variables, which tells how variables in the solution interact to have an effect in the fitness function. Variation that does not take this interaction into account may not effectively optimise the fitness function. The need to discover interaction has led to the development of the probabilistic approach to variation, where this interaction is used to build a probabilistic model of the solutions. The built model is then sampled to generate the offspring population.

Siddhartha Shakya
Business Modelling and Operational Transformation Practice, BT Innovate & Design, Ipswich, UK
e-mail: sid.shakya@bt.com

Roberto Santana
Departamento de Inteligencia Artificial, Universidad Politécnica de Madrid, Madrid, Spain
e-mail: roberto.santana@upm.es

S. Shakya and R. Santana (Eds.): Markov Networks in Evolutionary Computation, ALO 14, pp. 21–37.
springerlink.com

Algorithms using this approach to variation are known as Estimation of Distribution Algorithms (EDAs) [34, 45]. They are a class of EAs that extracts statistical information from the population of solutions and uses it to generate new solutions. Several results have shown [38, 52] that they are able to solve problems that are known to be hard for traditional Genetic Algorithms (GA) [18, 26]. An EDA maintains the selection and variation concepts of evolution. However, it replaces the crossover and mutation approach to variation in a traditional GA by building and sampling a probabilistic model of solutions. As such, the evolution process in an EDA can be seen as explicitly biased towards the significant patterns identified by the probabilistic models. This contrasts with the more implicit processing of important patterns in traditional GAs. The key motivation behind EDAs is to identify and exploit the linkage between variables in the solution to assist the evolution. However, there are two more factors, as noted by [34], that also motivated the researchers towards a probabilistic approach to evolution. Firstly, the performance of GA depends on the choice of parameters and design factors, such as different crossover and mutation operators, probabilities associated with crossover and mutation, population size and so on. Therefore, choosing an effective configuration for a GA can become an optimisation problem in itself [20]. This was one of the motivation for early EDAs, which try to minimise (or at least make it easy to set) the parameters for the algorithm. Second motivation for EDA arises from the theoretical analysis side of GA, which is an extremely difficult task. Several theories have been proposed to explain the GA evolution but with moderate success. It was assumed that with EDAs a better and more rigorous theoretical analysis of the evolutionary process could be achieved [34].

In this chapter, we review some of the popular EDAs. Our key focus will be on Markov network based EDAs. Section 2 describes the workflow of a typical EDA. Section 3 proposes a categorisation of EDA based on the type of Probabilistic Graphical Models (PGM) they use and lists some of the popular EDAs in each category. Section 4 discusses in more detail the EDAs based on Markov network. It also outlines some of the most recent works in the area of Markov network based EDAs.

2.2 Estimation of Distribution Algorithm

An EDA regards a solution, $x = \{x_1, x_2, .., x_n\}$, as a set of values taken by a set of variables, $X = \{X_1, X_2, .., X_n\}$. As shown in Figure 2.1, it starts by initialising a population of solutions, P. A set of promising solutions D is then selected from P, and is used to estimate a probabilistic model of X. The model is then sampled to generate the next population.

2.2.1 A Simple EDA Workflow

Let us give an example of a very simple EDA. Figure 2.2 shows an EDA that considers each variable as independent variables. As such, the joint probability can be

---

Estimation of Distribution Algorithm

1. Generate initial (parent) population P of size M
2. Select set D from P consisting of N solutions, where $N \leq M$
3. Estimate the probability distribution of variables in the solutions from D
4. Sample distribution to generate offspring, and replace parents
5. Go to step 2 until termination criteria are met

Fig. 2.1 The workflow of the general Estimation of Distribution Algorithm

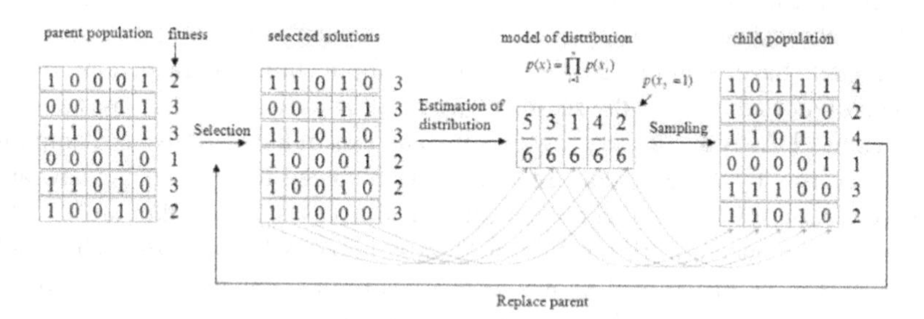

Fig. 2.2 A workflow of a simple EDA

simply modelled as the product of marginal probabilities of individual variables. Following the general EDA workflow, it starts by generating the set of parent population. It then evaluates each individual solution to get their fitness value. It then selects a set of good solution from the parent set (in this example, a single solution can be repeatedly selected). It then calculates the marginal probability for each variable $p(x_i)$. In this case the probability of each variable being 1, $p(x_i = 1)$, is calculated. From which it is easy to calculate $p(x_i = 0)$ as $p(x_i = 0) = 1 - p(x_i = 1)$. These marginal probabilities are then sampled to create new solutions. The new set of solutions replaces the old set and this process continues until a termination criterion is met. It can be noticed that the process is similar to that of GA, except the crossover and mutation is replaced with a probabilistic modelling and sampling process. This is a very simple algorithm, yet can be efficient in problems were a fast convergence to a solution is required.

2.2.2 *Probabilistic Graphical Models and EDAs*

The example shown in figure 2.2 is the simplest class of EDAs assuming each variable are independent. Although it can be effective in some problems, this representation may not always be the true representation of the variable interaction in the problem being addressed. In fact, it is very likely that in real life problems the variables are highly interrelated. As such these relationships have to be identified, and then properly modelled. For example, assuming that the two variables are mutually dependent to each other, we can have a joint probability model instead of the univariate probability model. This is where the probabilistic graphical models (PGM) [35] can be useful. As stated in previous chapter, PGMs are the efficient and effective tool to model the probabilistic relationship between variables[1]. It is obvious that the effectiveness of any EDAs very much depends on how well they can model this relationship, and therefore probabilistic modelling lies at the very heart of EDAs. It is for this reason PGMs are widely used in EDAs to represent probability distribution. Particularly, Directed Graphical Models (e.g. Bayesian networks) [48] have been widely studied and are well established as a useful approach for modelling the distribution in EDAs. Recent years have seen increasing interest in using Undirected Graphical Models (Markov networks / Markov Random Fields) [6, 37, 46] for EDAs. Nevertheless, in comparison with their counterparts, EDAs that use undirected models have been less studied and fewer applications to practical problems have been reported. One of the reasons why Markov networks (MN) are not well exploited in EDAs is due to very few existing analysis of the learning and sampling procedures used by these models. In addition to the need for introducing new and improved learning/sampling techniques, it is also necessary to further analyse the way these components work together in Markov network based EDAs. This book consolidates some of the well known techniques in this area and aims to provide practical information on building Markov network models for EDAs.

2.3 EDA Categorisation

Many literatures categorise EDA according to the complexity of the probability model they use.

1. Univariate EDA
2. Bivariate EDA
3. Multivariate EDA

Univariate EDAs are those that assume solution variables to be independent and therefore build univariate marginal models of the solution variables. Bivariate EDAs assumes that at most two variables are interdependent and therefore builds a bivariate model of probability distribution. Finally multivariate EDAs build probabilistic model consisting of multiple dependent variables.

Another popular EDA categorisation is based on the way the variables are represented in a solution.

1. Discrete EDA
2. Continuous EDA
3. Mixed EDA

Discrete EDAs assume each variable in the problem to be discrete. This includes many popular binary EDAs. Similarly continuous EDAs assume variables in the

[1] Chapter 1 introduces some of the popular PGMs.

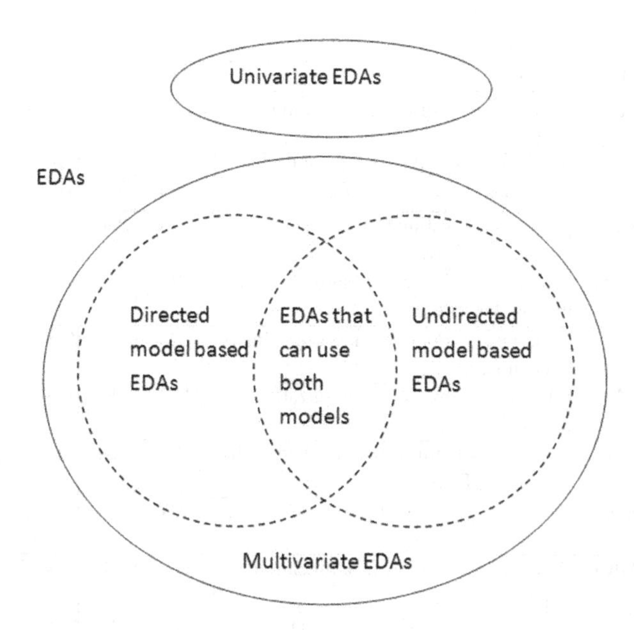

Fig. 2.3 PGM based EDA catagorisation

problem are continuous. They mainly exploit Gaussian models. Finally mixture models based EDAs can have mixture of both discrete and continuous variables in the problem.

As our focus is on Markov networks and the PGMs, it is natural for us to categorise EDAs according to the type of PGM they use. Figure 2.3 shows the proposed categorisation of EDAs according to the type of PGMs they use.

1. Directed Model based EDA
2. Undirected Model based EDA
3. Junction Tree based EDA (both directed and undirected model based)

We first divide EDAs into univariate and multivariate. Then we further divide multivariate EDAs into directed model based, undirected model based and common (both directed and undirected) model based. Next, we briefly list some of the popular EDAs in each of these categories. More extended list of EDAs is shown in Table 2.1. In section 2.4, we briefly describe the workflow of some of the popular Markov network based EDAs.

2.3.1 Univariate EDA

As name suggests, these are EDAs that do not assume any dependency between variables. As such each variable is independent and is modelled as univariate marginal probability. The earliest EDA of this class is Population Based Incremental Learning

Table 2.1 The list of popular EDAs

EDA	Description	PGM	Order	Rep	Ref
UMDA	Univariate Marginal Distribution algorithm	UM	Uni	Disc	[45]
PBIL	Population Based Incremental Learning	UM	Uni	Disc	[3]
CGA	Compact Genetic Algorithm	UM	Uni	Disc	[23]
DEUMd	Distribution estimation using Markov Networks (Univariate)	UM	Uni	Disc	[63]
UMDAc	Continuous Univariate Marginal Distribution Algorithm	UM	Uni	Cont	[31]
PBILc	Continuous Population Based Incremental Learning	UM	Uni	Cont	[58]
MIMIC	Mutual Information Maximization for input clustering	BN	Bi	Disc	[13]
COMIT	Combining Optimizers with Mutual Information Trees	BN	Bi	Disc	[5]
BMDA	Bivariate Marginal Distribution Algorithm	BN	Bi	Disc	[53]
MIMICc	Mutual Information Maximization for input clustering	GN	Bi	Cont	[30, 33]
BMDAc	Continuous Bivariate Marginal Distribution Algorithm	GN	Bi	Cont	[30, 33]
FDA	Factorised Distribution Algorithm	BN/MN	Multi	Disc	[44, 47]
ECGA	Extended Compact Genetic algorithm	BN/MN	Multi	Disc	[22]
EBNA	Estimation of Bayesian Network Algorithm	BN	Multi	Disc	[14]
LFDA	Learning Factorised Distribution Algorithm	BN	Multi	Disc	[43]
BOA/hBOA	Bayesian optimisation algorithm	BN	Multi	Disc	[49, 51]
MN-FDA	Markov Network Factorised Distribution algorithm	BN/MN	Multi	Disc	[54]
MN-EDA	Markov Network Estimation of distribution algorithm	MN	Multi	Disc	[55]
DEUM	Distribution Estimation using Markov Networks	MN	Multi	Disc	[59, 60]
MARLEDA	Markovian Learning Estimation of Distribution Algorithm	MN	Multi	Disc	[2]
MOA	Markovianity Optimisation Algorithm	MN	Multi	Disc	[65, 66]
IDEA	Iterated Density Estimation Algorithm	GN	Multi	Cont	[7, 8]
EMNA	Estimation of Multivariate Normal Algorithm	GN	Multi	Cont	[31]
EGNA	Estimation of Gaussian Network Algorithm	GN	Multi	Cont	[31]
BOAc	Real-code Bayesian Optimisation Algorithm	BN	Multi	Cont	[1]
MBOA	Multi-objective Bayesian Optimisation Algorithm	BN	Multi	Cont	[27, 28]
MIDEA	Multi-objective Iterated Density Estimation Algorithm	BN	Multi	Cont	[9]

algorithm (PBIL) [3, 4]. It is a binary EDA, motivated by the idea of combining GAs with Competitive Learning used in training Artificial Neural Networks. PBIL starts with initialisation of a probability vector. It then iteratively updates and samples the probability vector to generate better solutions.

Another popular univariate EDA, Univariate Marginal Distribution Algorithm (UMDA) proposed by [45] is another of the early EDAs. Different variants of UMDA have been proposed, and the mathematical analysis of their workflows has been carried out [19, 41, 44]. Its workflow is similar to that of PBIL, except it does not have an additional parameter (called learning rate) that controls the speed in which the probability distribution converges. As such it can be seen as simpler version of PBIL.

Compact Genetic Algorithm (cGA) is another of the univariate EDAs, proposed by [23]. It is motivated by the previous work done in the field of random walk model [21], and also assumes no interaction between variables in solution. cGA also maintains a probability vector as in PBIL. However, unlike PBIL, cGA samples only two solutions at a time, compares their fitness, and uses allele value of the winning solution (i.e solution with the highest fitness) to update the probability vector, leaving the probability vector unchanged in the position, where winning and losing solution contains the same value. The process continues until the probability vector converges.

Distribution Estimation using Markov network with Direct sampling ($DEUM_d$) [61–63] is another univariate EDA, which is motivated by the concept of energy modelling in Markov networks. It is substantially different from the other three univariate EDAs. It maintains a vector of univariate marginal probabilities but the probabilities are calculated using a log-linear model of fitness function that is approximated from the population of solution. More on DEUMd is described in chapter 4.

Although Univariate EDAs use simple univariate models, they have been shown to work very well in practice. Also their workflow is much simpler than other multivariate EDAs and other traditional EAs, converging quickly and efficiently to a solution. Therefore, they are of prominent importance to EA community, and should be given serious consideration when designing real world optimisation models.

2.3.2 Multivariate EDA

These are EDAs that take into account the dependency between variables and builds a multivariate probability model of the solution. These models are then sampled to generate new populations. These multivariate probability models are represented as PGMs. The decision to use a specific type of PGM depends on the type of the relationship between variables. If problem variables have causal relationship, directed PGM is preferred. Similarly, if the problem variables are mutually dependent, undirected PGM is preferred.

2.3.2.1 Directed Model Based EDA

Directed PGM (Bayesian network) based EDAs assume causality in variables interaction and build a acyclic conditional probability model. The key aspect of Bayesian network based EDA is to find the correct causal interaction between variables. Once this is found, the model building, parameter estimation and sampling is relatively easier. That is mainly because directed acyclic graphs can be sampled very efficiently, simply by following the parent child relationship as modelled by the structure of the network.

Most of the early multivariate EDAs uses Bayesian network approach to modelling and sampling. Some restrict themselves to pair-wise interaction, such as Combining Optimizers with Mutual Information Trees (COMIT)[5], Mutual Information Maximization for Input Clustering (MIMIC) [13] and Bivariate Marginal Distribution Algorithm (BMDA) [53], and some allow fully multivariate dependencies, such as Bayesian Optimisation Algorithm (BOA) [51], hierarchical Bayesian Optimisation Algorithm (hBOA) [50], Estimation of Bayesian network Algorithm (EBNA) [14, 32] and Learning Factorised Distribution Algorithm (LFDA) [43]. Recently, EDAs that employ dependency networks [25] (directed networks that allow the existence of cycles) have been also introduced [15, 16]. The work presented in [34, 49] provides a good review of Bayesian network based EDAs and their structure learning and sampling processes.

2.3.2.2 Undirected Model Based EDA

These are EDAs that use undirected PGMs to represent interaction between variables. As such, they can model mutual dependencies between variables more naturally. Because of the nature of two-way mutual dependency between problem variables, global property (the joint probability distribution) in Markov network do not directly relate to the local conditional probabilities (See chapter 1 on PGM for more on global and local properties of Markov network). Therefore, the EDAs in this category can be further divided into the following two sub-categories.

1. Global property based
2. Local property based

Most of the early Markov network based EDAs fall into the category of global Markov property based EDAs. They directly model the joint probability distribution of Markov network and sample from it. Distribution Estimation using Markov networks (DEUM) [59, 60], Markov network Estimation of Distribution Algorithm (MN-EDA) [55] and Markov network Factorized Distributing Algorithm (MN-FDA) [54] are classed in this category. Some recent work in Markov network base EDA focuses on directly using conditional probabilities as modelled by the local Markov property, the Markovianity. They are Markovian Optimisation Algorithm (MOA) [65, 66] and Markovian Learning Estimation of Distribution Algorithm (MARLEDA) [2]. In section 2.4, we provide case for using Markov networks in EDAs and briefly review both global and local Markov property based EDAs. There is however a third class of EDA, which can also be categorised as Markov network based EDAs. Extended Compact Genetic Algorithm (ECGA) [22], an extension to cGA, falls in this category. ECGA models joint probability as the product of marginal probabilities of disjoint sets of variables. It samples these marginal probabilities to get the new solution. Each of these disjoint sets can be taken as fully connected Markov networks. Therefore we categorise ECGA as a special case of Markov network based EDAs.

2.3.2.3 Common Model Based EDA

These are multivariate EDAs whose probability model can be represented as both Markov network and Bayesian network. They use junction trees as their probability model, also widely known as triangulated or chordal graphical model [35, 36]. Junction trees are undirected graphs which satisfy the running intersection properties (as described in chapter 1) and therefore can also be modelled as directed graphs. The PLS approach to sampling of Bayesian network, therefore, also applies to these EDAs. Factorised Distribution Algorithm (FDA) [44], one of the early multivariate EDAs, use junction tree approach to estimating and sampling the distribution. FDA assumes that an undirected problem structure is known in advance and then translates this structure into a junction tree. An FDA able to learn a junction tree from the data was introduced in [47]. Another EDA, Linkage Detection Factorization Algorithm (LDFA), proposed by [70] also exploit junction trees.

2.4 Estimation of Distribution Algorithms with Markov Networks

Now that we have categorised EDAs by their PGMs, let us look into MN based EDAs in more detail, and understand some of their key properties.

2.4.1 Motivation

As stated earlier, the research on EDA that learns higher order probability distributions have been primarily focused on Bayesian network based EDAs. However, in recent years, we have seen an increasing interest in the use of Markov networks in EDAs. Learning the accurate causal model forms the bottle neck for Bayesian network based approach. The structure learning process can be complicated, and may not naturally align with the true interaction between variables. One of the typical examples of such problem is Ising spin glasses [29], where perfect Bayesian network model is difficult to achieve [24]. In problems, where the interactions between the variables form loops and are difficult to capture with a Bayesian network, Markov network is a natural alternative. It does not require causality principle to be satisfied, and can naturally represent mutual two-way dependencies. Many problems in real life can be seen as an instance of having such mutual relationships. There is however a trade-off. The ability to correctly model the natural dependencies in Markov network comes with an increased cost in sampling. The mutual dependency encoded in

Markov network does not naturally provide the order in which the variables can be sampled. As such, sampling forms the bottleneck for Markov network. Some random walk sampling methods, such as Markov chain Monte-Carlo samplings, have been investigated and have been shown to work effectively in Markov networks based EDAs. Figure 2.4 shows the workflow of a general EDA based on Markov networks.

EDA based on Markov networks

1. Generate initial (parent) population P of size M
2. Select set D from P consisting of N solutions, where $N \leq M$
3. Estimate an undirected graph capturing the interactions between variables from D
4. Build a Markov network model and estimate the model parameters
5. Sample Markov network to generate offspring, and replace parents
6. Go to step 2 until termination criteria are met

Fig. 2.4 The workflow of general EDA based on Markov networks

2.4.2 History

One of the early works on the use of Markov network in EA, was focused on relating GA fitness function with the energy function of Markov networks [10]. The idea was to model the fitness function as an instance of Markov networks that could be used to help understand the evolution process in GAs. Particularly, the model was used for two purposes. A) To predict the fitness of the next GA generation and B) to predict the optimal crossover and mutation point for a solution. [61] extended this fitness modelling approach to the EDA domain. The idea was to use it as a model of distribution within an EDA. The outcome was one of the early Markov network based EDA, known as DEUM. Independent to this, the work of [54] also focused on approximating the joint probability distribution in terms of the cliques in Markov networks. The outcome of this was a Markov network EDA based on Kikuchi approximation (MN-EDA) [55]. Let us briefly review these and other recent approaches to modelling distribution in EDAs with Markov networks.

2.4.3 Review of Popular EDAs with Markov Networks

Two EDAs DEUM [59] and MN-EDA [55] exploit global property of Markov networks. More precisely, they factorise the joint probability distribution of a Markov network in terms of cliques in the undirected graph and sample it to generate new solutions. There are other more recent Markov network based EDAs that do not factorise the joint probability distribution. Instead they directly sample from local conditional probabilities modelled by the neighbourhood structure of the Markov network. Next we briefly review some of the key principles of these Markov network based EDAs.

DEUM

DEUM is a family of Markov network based EDAs that builds a model of fitness function in terms of the cliques in the undirected graph and factorises the joint probability as a Gibbs distribution. The parameters of the fitness model is then estimated from the population of solutions and Markov chain Monte Carlo simulations, including Gibbs sampler [17] and Metropolis sampler [40], are used to sample new solutions. Several variants of DEUM have been proposed and are found to perform well in comparison to other EDAs of their class in a range of different test problems, including Ising Spin Glass and SAT [61, 63, 64]. Recent improvements to DEUM includes the use of structural learning methods to learn the models [11, 12, 67]. Chapter 4 of this book gives a detail introduction to DEUM algorithms.

MN-EDA and MN-FDA

MN-FDA [54] and MN-EDA [55] are based on the idea of making an approximation to the joint probability distribution in terms of cliques in the undirected graph. MN-EDA does so by means of the so-called Kikuchi approximation of the joint probability distribution inspired in the application of free energy based approximations in statistical physics [71]. The structure of the model can be given to the algorithm or learned from data using statistical tests. The algorithm uses Gibbs sampler to sample the new solutions. To approximate the joint probability, MN-FDA constructs a junction graph [55] from the undirected structure. The junction graph is then sampled using PLS to generate new solutions. Chapter 5 of this book is dedicated to these algorithms.

MOA, MARLEDA and MN-GIBBS

Some recent Markov network based EDAs also exploit the local Markov property of Markov networks. In other words, they directly sample from the conditional probabilities encoded by the neighbourhood relationship in the undirected graph. They do not model the joint distribution of the solutions and therefore are much simpler than other global Markov property based EDAs. Furthermore, in addition to gain in efficiency, they avoid the numerical operations associated to the computation of cliques and their potentials. MOA [65, 66], MARLEDA [2] and a general class of Markov network algorithms (MN-GIBBS) [42] fall in this category. MARLEDA does structural learning of the Markov network by means of Pearson X^2 statistical tests in an approach similar to the ones used in [12, 55]. MN-GIBBs computes the Markov network from the factor graph associated to the fitness function, i.e. no structural learning is done, although an approach for learning constrained Markov networks from data is discussed in [42]. On the other hand, MOA uses a mutual information based approach to learning the undirected structure, and also uses an advanced Gibbs sampler algorithm to sample from the model. This temperature based annealing schedule in Gibbs sampling gives MOA an extra edge, allowing to explicitly balance the exploitation and exploitation of the search space. Chapter 3 of this book describes in detail the MOA approach and its structure learning and sampling techniques.

2.4.4 Current Research on EDA Based on Markov Networks

Very recent research on Markov network based EDAs have expanded over a variety of areas. Some of these areas such as real-world applications, fitness-modelling and application to continuous optimization problems are covered in different chapters of this book. In this section we briefly review other questions that are relevant for the development of Markov network based EDAs and where on-going research is pursued. New approaches for learning the Markov network structures include the application of regularization techniques [39] and the design of more efficient learning methods based on hybrid structure-discovery techniques [57]. In [39, 68], the task of selecting the proper structure of the Markov network is addressed by using l1-regularized logistic regression techniques. The idea is to use regularization to find a trade-off between the complexity of the learned models and their accuracy. The algorithm is tested in the context of DEUM showing a good performance for several difficult functions. Another approach is based on the use of a hybrid approach (score+search and independence tests) to learn more accurate Markov network structures. In [57], the Independence Based Maximum a Posteriori (IBMAP) approach for robust learning of Markov networks is presented and tested in the context of evolutionary optimization using MOA.

Current research also emphasizes the importance of flexible and robust implementations of Markov-network based EDAs [56, 69]. Research on EDA software implementations is very important since the design of general EDA programs, able to adjust to diverse real-world problems, is not a straightforward task. One novel approach to the implementations of Markov-network EDAs consider the use of modular architectures in which the different EDA components can be combined or replaced by the user according to the characteristics of the optimization problem. These approaches easily allow the implementation of hybrid schemes in which EDAs are combined with local optimization methods, GAs and other optimization algorithms. There are many areas where research on Markov network EDAs is required. One of the open questions is related to the conception of efficient and accurate sampling methods. Current Gibbs sampling based algorithms experience some difficulties to scale well when the number of variables or the cardinality of the problem variables is increased. More robust version of sampling algorithms would be a valuable input to this approach. Also, some work is needed in the way the structure of the network is learnt. Although, it does not require finding causality relationship, it can still get complicated with the size of the problem and the order of variable interaction. There are other areas of research related to different instances of Markov network based EDAs, such as fitness modelling, energy approximation, and model fittings. They will be discussed in subsequent chapters of this book.

Acknowledgements. This work has been partially supported by the Saiotek and Research Groups 2007-2012 (IT-242-07) programs (Basque Government), TIN2010-14931 and Consolider Ingenio 2010 - CSD 2007 - 00018 projects (Spanish Ministry of Science and Innovation).

References

1. Ahn, C.W., Kim, K.P., Ramakrishna, R.S.: A Memory-Efficient Elitist Genetic Algorithm, pp. 552–559. Springer (2004)
2. Alden, M.A.: MARLEDA: Effective Distribution Estimation Through Markov Random Fields. PhD thesis. Faculty of the Graduate Schoool. University of Texas at Austin, USA (December 2007)
3. Baluja, S.: Population-based incremental learning: A method for integrating genetic search based function optimization and competitive learning. Technical Report CMU-CS-94-163, Pittsburgh, PA (1994)
4. Baluja, S.: An empirical comparison of seven iterative and evolutionary function optimization heuristics. Technical Report CMU-CS-95-193. Carnegie Mellon University (1995)
5. Baluja, S., Davies, S.: Using optimal dependency-trees for combinatorial optimization: Learning the structure of the search space. In: Proceedings of the 14th International Conference on Machine Learning, pp. 30–38. Morgan Kaufmann (1997)
6. Besag, J.: Spatial interactions and the statistical analysis of lattice systems. Journal of the Royal Statistical Society B-36, 192–236 (1974)
7. Bosman, P.A.: Design and Application of Iterated Density-Estimation Evolutionary Algorithms. PhD thesis. Universiteit Utrecht. Utrecht, The Netherlands (2003)
8. Bosman, P.A., Thierens, D.: Expanding from Discrete to Continuous Estimation of Distribution Algorithms: The IDEA. In: Deb, K., Rudolph, G., Lutton, E., Merelo, J.J., Schoenauer, M., Schwefel, H.-P., Yao, X. (eds.) PPSN 2000. LNCS, vol. 1917, pp. 767–776. Springer, Heidelberg (2000)
9. Bosman, P.A., Thierens, D.: Multi-objective optimization with diversity preserving mixture-based iterated density estimation evolutionary algorithms. International Journal of Approximate Reasoning 31(3), 259–289 (2002)
10. Brown, D.F., Garmendia-Doval, A.B., McCall, J.A.W.: Markov Random Field Modelling of Royal Road Genetic Algorithms. In: Collet, P., et al. (eds.) EA 2001. LNCS, vol. 2310, pp. 65–78. Springer, Heidelberg (2002)
11. Brownlee, A.E.I.: Multivariate Markov networks for fitness modelling in an estimation of distribution algorithm. PhD thesis. The Robert Gordon University. School of Computing, Aberdeen, UK (2009)
12. Brownlee, A.E.I., McCall, J., Shakya, S.K., Zhang, Q.: Structure learning and optimisation in a Markov-network based estimation of distribution algorithm. In: Proceedings of the 2009 Congress on Evolutionary Computation CEC-2009, pp. 447–454. IEEE Press, Norway (2009)

13. de Bonet, J.S., Isbell Jr., C.L., Viola, P.: MIMIC: Finding optima by estimating probability densities. In: Mozer, M.C., Jordan, M.I., Petsche, T. (eds.) Advances in Neural Information Processing Systems, vol. 9. The MIT Press (1997)
14. Etxeberria, R., Larrañaga, P.: Global optimization using Bayesian networks. In: Ochoa, A., Soto, M.R., Santana, R. (eds.) Proceedings of the Second Symposium on Artificial Intelligence (CIMAF 1999), Havana, Cuba, pp. 151–173 (1999)
15. Gámez, J.A., Mateo, J.L., Puerta, J.M.: EDNA: Estimation of Dependency Networks Algorithm. In: Mira, J., Álvarez, J.R. (eds.) IWINAC 2007. LNCS, vol. 4527, pp. 427–436. Springer, Heidelberg (2007)
16. Gámez, J.A., Mateo, J.L., Puerta, J.M.: Improved EDNA estimation of dependency networks algorithm using combining function with bivariate probability distributions. In: Proceedings of the 10th Annual Conference on Genetic and Evolutionary Computation, GECCO 2008, pp. 407–414. ACM, New York (2008)
17. Geman, S., Geman, D.: Stochastic relaxation, Gibbs distributions and the Bayesian restoration of images. In: Fischler, M.A., Firschein, O. (eds.) Readings in Computer Vision: Issues, Problems, Principles, and Paradigms, pp. 564–584. Kaufmann, Los Altos (1987)
18. Goldberg, D.: Genetic Algorithms in Search, Optimization, and Machine Learning. Addison-Wesley (1989)
19. González, C., Rodríguez, J.D., Lozano, J., Larrañaga, P.: Analysis of the Univariate Marginal Distribution Algorithm modeled by Markov chains. In: Mira, J., Álvarez, J.R. (eds.) IWANN 2003. LNCS, vol. 2686, pp. 510–517. Springer, Heidelberg (2003)
20. Grefenstette, J.J.: Optimization of control parameters for genetic algorithms. IEEE Transactions on Systems, Man, and Cybernetics 16, 122–128 (1986)
21. Harik, Cantu-Paz, Goldberg, Miller: The gambler's ruin problem, genetic algorithms, and the sizing of populations. In: IEEECEP: Proceedings of The IEEE Conference on Evolutionary Computation, IEEE World Congress on Computational Intelligence (1997)
22. Harik, G.: Linkage learning via probabilistic modeling in the ECGA. Technical Report IlliGAL Report No. 99010. University of Illinois at Urbana-Champaign (1999)
23. Harik, G.R., Lobo, F.G., Goldberg, D.E.: The compact genetic algorithm. IEEE-EC 3(4), 287 (1999)
24. Hauschild, M., Pelikan, M., Lima, C., Sastry, K.: Analyzing probabilistic models in hierarchical BOA on traps and spin glasses. In: Thierens, D., et al. (eds.) Proceedings of the Genetic and Evolutionary Computation Conference, GECCO 2007, vol. I, pp. 523–530. ACM Press, London (2007)
25. Heckerman, D., Chickering, D.M., Meek, C., Rounthwaite, R., Kadie, C.M.: Dependency networks for inference, collaborative filtering, and data visualization. Journal of Machine Learning Research 1, 49–75 (2000)
26. Holland, J.H.: Adaptation in Natural and Artificial Systems. University of Michigan Press, Ann Arbor (1975)
27. Khan, N.: Bayesian optimization algorithms for multi-objective and hierarchically difficult problems. Master's thesis. University of Illinois at Urbana-Champaign, Illinois Genetic Algorithms Laboratory, Urbana, IL (2003)
28. Khan, N., Goldberg, D.E., Pelikan, M.: Multi-objective Bayesian optimization algorithm. IlliGAL Report No. 2002009. University of Illinois at Urbana-Champaign, Illinois Genetic Algorithms Laboratory, Urbana, IL (2002)
29. Kindermann, R., Snell, J.L.: Markov Random Fields and Their Applications. AMS (1980)

30. Larrañaga, P., Etxeberria, R., Lozano, J., Peña, J.: Optimization by learning and simulation of Bayesian and Gaussian networks. Technical Report EHU-KZAA-IK-4/99. University of the Basque Country (1999)
31. Larrañaga, P., Etxeberria, R., Lozano, J.A., Peña, J.M.: Optimization by learning and simulation of Bayesian and Gaussian networks. Technical Report EHU-KZAA-IK-4/99. Department of Computer Science and Artificial Intelligence. University of the Basque Country (1999)
32. Larrañaga, P., Etxeberria, R., Lozano, J.A., Peña, J.M.: Combinatorial optimization by learning and simulation of Bayesian networks. In: Proceedings of the Sixteenth Conference on Uncertainty in Artificial Intelligence, Stanford, pp. 343–352 (2000)
33. Larrañaga, P., Etxeberria, R., Lozano, J.A., Peña, J.M.: Optimization in continuous domains by learning and simulation of Gaussian networks. In: Wu, A.S. (ed.) Proceedings of the 2000 Genetic and Evolutionary Computation Conference Workshop Program, pp. 201–204 (2000)
34. Larrañaga, P., Lozano, J.A.: Estimation of Distribution Algorithms: A New Tool for Evolutionary Computation. Kluwer Academic Publishers (2002)
35. Lauritzen, S.L.: Graphical Models. Oxford University Press (1996)
36. Lauritzen, S.L., Spiegelhalter, D.J.: Local computations with probabilities on graphical structures and their application to expert systems. Journal of the Royal Statistical Society B 50, 157–224 (1988)
37. Li, S.Z.: Markov Random Field modeling in computer vision. Springer (1995)
38. Lozano, J.A., Larrañaga, P., Inza, I., Bengoetxea, E. (eds.): Towards a New Evolutionary Computation: Advances on Estimation of Distribution Algorithms. Springer (2006)
39. Malagó, L., Matteo, M., Gabriele, V.: Introducing l1-regularized logistic regression in Markov networks based EDAs. In: Proceedings of the 2011 Congress on Evolutionary Computation CEC 2011, pp. 1581–1588. IEEE (2011)
40. Metropolis, N.: Equations of state calculations by fast computational machine. Journal of Chemical Physics 21, 1087–1091 (1953)
41. Mühlenbein, H.: The equation for response to selection and its use for prediction. Evolutionary Computation 5(3), 303–346 (1998)
42. Mühlenbein, H.: Convergence of estimation of distribution algorithms (2009) (submmited for publication)
43. Mühlenbein, H., Mahnig, T.: FDA - A scalable evolutionary algorithm for the optimization of additively decomposed functions. Evolutionary Computation 7(4), 353–376 (1999)
44. Mühlenbein, H., Mahnig, T., Ochoa, A.: Schemata, distributions and graphical models in evolutionary optimization. Journal of Heuristics 5(2), 215–247 (1999)
45. Mühlenbein, H., Paaß, G.: From Recombination of Genes to the Estimation of Distributions: I. Binary Parameters. In: Ebeling, W., Rechenberg, I., Voigt, H.-M., Schwefel, H.-P. (eds.) PPSN 1996. LNCS, vol. 1141, pp. 178–187. Springer, Heidelberg (1996)
46. Murray, I., Ghahramani, Z.: Bayesian Learning in Undirected Graphical Models: Approximate MCMC algorithms. In: Twentieth Conference on Uncertainty in Artificial Intelligence (UAI 2004), Banff, Canada, July 8-11, pp. 392–399 (2004)
47. Ochoa, A., Soto, M.R., Santana, R., Madera, J., Jorge, N.: The factorized distribution algorithm and the junction tree: A learning perspective. In: Ochoa, A., Soto, M.R., Santana, R. (eds.) Proceedings of the Second Symposium on Artificial Intelligence (CIMAF 1999), pp. 368–377. Editorial Academia, Havana (1999)
48. Pearl, J.: Probabilistic Reasoning in Intelligent Systems: Networks of Plausible Inference. Morgan Kaufmann, San Mateo (1988)

49. Pelikan, M.: Hierarchical Bayesian optimization algorithm: Toward a new generation of evolutionary algorithms. Springer (2005)
50. Pelikan, M., Goldberg, D.E.: Hierarchical problem solving by the Bayesian optimization algorithm. IlliGAL Report No. 2000002. Illinois Genetic Algorithms Laboratory. University of Illinois at Urbana-Champaign, Urbana, IL (2000)
51. Pelikan, M., Goldberg, D.E., Cantú-Paz, E.: BOA: The Bayesian Optimization Algorithm. In: Banzhaf, W., et al. (eds.) Proceedings of the Genetic and Evolutionary Computation Conference GECCO 1999, vol. I, pp. 525–532. Morgan Kaufmann Publishers, San Francisco (1999)
52. Pelikan, M., Goldberg, D.E., Lobo, F.: A survey of optimization by building and using probabilistic models. Computational Optimization and Applications 21(1), 5–20 (2002)
53. Pelikan, M., Mühlenbein, H.: The bivariate marginal distribution algorithm. In: Roy, R., Furuhashi, T., Chawdhry, P.K. (eds.) Advances in Soft Computing - Engineering Design and Manufacturing, pp. 521–535. Springer, London (1999)
54. Santana, R.: A Markov Network Based Factorized Distribution Algorithm for Optimization. In: Lavrač, N., Gamberger, D., Todorovski, L., Blockeel, H. (eds.) ECML 2003. LNCS (LNAI), vol. 2837, pp. 337–348. Springer, Heidelberg (2003)
55. Santana, R.: Estimation of distribution algorithms with Kikuchi approximations. Evolutionary Computation 13(1), 67–97 (2005)
56. Santana, R., Bielza, C., Larrañaga, P., Lozano, J.A., Echegoyen, C., Mendiburu, A., Armañanzas, R., Shakya, S.: Mateda-2.0: A MATLAB package for the implementation and analysis of estimation of distribution algorithms. Journal of Statistical Software 35(7), 1–30 (2010)
57. Schlüter, F., Bromberg, F.: Independence-based MAP for Markov networks structure discovery. In: Proceedings of the 23rd IEEE International Conference on Tools with Artificial Intelligence (2011) (in press)
58. Sebag, M., Ducoulombier, A.: Extending Population-Based Incremental Learning to Continuous Search Spaces. In: Eiben, A.E., Bäck, T., Schoenauer, M., Schwefel, H.-P. (eds.) PPSN 1998. LNCS, vol. 1498, pp. 418–427. Springer, Heidelberg (1998)
59. Shakya, S.: DEUM: A Framework for an Estimation of Distribution Algorithm based on Markov Random Fields. PhD thesis. The Robert Gordon University, Aberdeen, UK (April 2006)
60. Shakya, S., McCall, J.: Optimisation by Estimation of Distribution with DEUM framework based on Markov Random Fields. International Journal of Automation and Computing 4, 262–272 (2007)
61. Shakya, S., McCall, J., Brown, D.: Updating the probability vector using MRF technique for a univariate EDA. In: Onaindia, E., Staab, S. (eds.) Proceedings of the Second Starting AI Researchers' Symposium. Frontiers in Artificial Intelligence and Applications, vol. 109, pp. 15–25. IOS press, Valencia (2004)
62. Shakya, S., McCall, J., Brown, D.: Estimating the distribution in an EDA. In: Ribeiro, B., Albrechet, R.F., Dobnikar, A., Pearson, D.W., Steele, N.C. (eds.) Proceedings of the International Conference on Adaptive and Natural Computing Algorithms (ICANNGA 2005), Coimbra, Portugal, pp. 202–205. Springer, Heidelberg (2005)
63. Shakya, S., McCall, J., Brown, D.: Using a Markov Network Model in a Univariate EDA: An Emperical Cost-Benefit Analysis. In: Proceedings of Genetic and Evolutionary Computation Conference (GECCO 2005), pp. 727–734. ACM, Washington, D.C. (2005)
64. Shakya, S., McCall, J., Brown, D.: Solving the Ising spin glass problem using a bivariate EDA based on Markov Random Fields. In: Proceedings of IEEE Congress on Evolutionary Computation (IEEE CEC 2006), pp. 3250–3257. IEEE press, Vancouver (2006)

65. Shakya, S., Santana, R.: An EDA based on local Markov property and Gibbs sampling. In: Keijzer, M. (ed.) Proceedings of the 2008 Genetic and Evolutionary Computation Conference (GECCO), pp. 475–476. ACM, New York (2008)
66. Shakya, S., Santana, R.: A markovianity based optimisation algorithm. Genetic Programming and Evolvable Machines (2011) (in press)
67. Shakya, S.K., Brownlee, A.E.I., McCall, J., Fournier, W., Owusu, G.: A fully multivariate DEUM algorithm. In: Proceedings of the 2009 Congress on Evolutionary Computation, CEC 2009, pp. 479–486. IEEE Press, Norway (2009)
68. Valentini, G.: A novel approach to model selection in distribution estimation using Markov networks. PhD thesis, Milan, Italy (2011)
69. Valentini, G., Malago, L., Matteucci, M.: Evoptool: An extensible toolkit for evolutionary optimization algorithms comparison. In: Proceedings of the 2010 IEEE Congress on Evolutionary Computation (CEC 2010), pp. 1–8. IEEE (2010)
70. Wright, A.H., Pulavarty, S.: Estimation of distribution algorithm based on linkage discovery and factorization. In: Proceedings of Genetic and Evolutionary Computation Conference (GECCO 2005), pp. 695–703. ACM, Washington, D.C. (2005)
71. Yedidia, J.S., Freeman, W.T., Weiss, Y.: Constructing free-energy approximations and generalized belief propagation algorithms. IEEE Transactions on Information Theory 51, 2282–2312 (2005)

Chapter 3
MOA - Markovian Optimisation Algorithm

Siddhartha Shakya and Roberto Santana

Abstract. In this chapter we describe Markovian Optimisation Algorithm (MOA), one of the recent developments in MN based EDA. It uses the local Markov property to model the dependency and directly sample from it without needing to approximate a complex join probability distribution model. MOA has a much simpler workflow in comparison to its global property based counter parts, since expensive processes to finding cliques, and building and estimating clique potential functions are avoided. The chapter is intended as an introductory chapter, and describes the motivation and the workflow of MOA. It also reviews some of the results obtained with it.

3.1 Introduction

Estimation of distribution algorithms [11] [18] are well established branch of Evolutionary computation [16]. They replace crossover and mutation approach to variation in traditional GAs [7] by probabilistic modelling and sampling of solution variables. Both Bayesian Networks (BN)[22] and Markov Networks (MN)[3][20] have been used for modelling the distribution in EDAs [24][5][10][31][27]. Here we focus on the later, the MN approach to modelling the distribution. Most of the early research on EDAs with MN mainly focused on exploiting its global property, the joint probability distribution (JPD) model. The key idea was to factorise the joint probability in terms of the cliques. The factorised JPD model was then sampled to

Siddhartha Shakya
Business Modelling and Operational Transformation Practice, BT Innovate & Design, Ipswich, UK
e-mail: sid.shakya@bt.com

Roberto Santana
Intelligent Systems Group. Faculty of Informatics,
University of the Basque Country (UPV/EHU), San-Sebastian, Spain
e-mail: roberto.santana@ehu.es

S. Shakya and R. Santana (Eds.): Markov Networks in Evolutionary Computation, ALO 14, pp. 39–53.
springerlink.com

get the new solution. In this chapter, we describe a recent development in EDAs with MN - an EDA that does not attempt to model the joint probability distribution, instead it exploits the local property of MNs, the conditional probability distribution based on neighbourhood relationship in an undirected graph. The algorithm is known as Markovian Optimisation Algorithm (MOA), sometimes also referred to as Markov Optimisation algorithm [28], Markovianity based Optimisation Algorithm [35]. The chapter is intended as an introductory chapter and describes the motivation and the workflow of MOA. It also presents some interesting results on the performance of MOA, which show that it can solve problems with complex interaction between variables and that its performance compares well with other multivariate EDAs.

The use of local Markov property in EDA was discussed in [31], where it was highlighted that the conditional probabilities encoded by a MN structure could be directly sampled without a need to modelling the joint probability distribution. The early description of MOA was published in [35], and more detail on it was covered in [37]. MOA incorporates features that have been independently employed in previous implementations of EDAs based on Markov models, but have not been used together. The resulting algorithm is qualitatively different to its predecessors. It does structural learning of the probabilistic model from the data but it can also take advantage of a-priori structural information in a straightforward way. Complex approximations to the joint probability distribution are avoided and the temperature parameter is included to balance the exploration and exploitation of the search space. The use of Gibbs sampling remains as a key component, allowing MOA to deal with interactions represented by cycles. [1] also proposed an EDA using similar concept of exploiting local MN property, called Markovian Learning Estimation of Distribution Algorithm (MARLEDA). Both MOA and MARLEDA have similar workflow, however they use different approaches to structure learning. MARLEDA uses χ^2 statistics, where as MOA uses a mutual information based approach. Also MOA uses an advanced Gibbs sampler algorithm for sampling the new solutions, which exploits the temperature parameter to balance the exploration and exploitation of the search space.

The outline of the chapter is as follows. Section 2 gives background on the local and global Markov properties in probabilistic graphical models. Section 3 reviews several Markov network based EDAs that make use of the global Markov properties. Section 4 presents a detailed description of the MOA. Section 5 presents some interesting results on the performance of MOA on several test functions. It highlights the importance of temperature based Gibbs sampling method in MOA. Finally, section 6 concludes the chapter.

3.2 Local and Global Properties in Graphical Models

Let us start this section by briefly reviewing how a solution is defined in EDAs. An EDA regards a solution, $x = \{x_1, x_2, .., x_n\}$, as a set of values taken by a set of variables, $X = \{X_1, X_2, ..., X_n\}$. EDAs begin by initialising a population of solutions, P. A set of promising solutions D is then selected from P, and is used to estimate a probabilistic model of X. The model is then sampled to generate the next population. Estimation of the probability distribution lies in the very heart of an EDA, largely dictating its effectiveness. This is where probabilistic graphical models [12] are useful.

As described in chapter 1, a PGM is characterised in terms of the joint probability of the nodes in the graph by its global property, the joint probability distribution, $p(x) = p(x_1, x_2, .., x_n)$. It is also characterised by the probabilities of individual nodes in the graph by it local property, the conditional probability of a variable given the rest of the variables in the graph $p(x_i|x - \{x_i\})$. The two key categories of PGMs are

1. Directed models (Bayesian networks)
2. Undirected models (Markov networks / Markov Random Fields)

In terms of BN, its global property, $p(x)$, is derived from its simplified local properties $p(x_i|\Pi_i)$, as

$$p(x) = \prod_{i=1}^{n} p(x_i|\Pi_i) \tag{3.1}$$

Where each $p(x_i|\Pi_i)$ is the set of probabilities associated with a variable $X_i = x_i$ given it's parent variables Π_i, as encoded in a directed acyclic graph. Therefore no distinction is made between global and local properties while categorising a BN.

In case of MNs however, the global property is not directly linked to its local property. The joint probability distribution is mainly defined in terms of cliques in the undirected graph as

$$p(x) = \frac{1}{Z} \prod_{i=1}^{m} \psi_i(c_i) \tag{3.2}$$

Where $\psi_i(c_i)$ (or more precisely $\psi_i(C_i = c_i)$) is a *potential function* on clique $C_i \in X$, m is the number of cliques in the structure G. $Z = \sum_{x \in \Omega} \prod_{i=1}^{m} \psi_i(c_i)$ is the normalising constant known as the *partition function* which ensures that $\sum_{x \in \Omega} p(x) = 1$. Here, Ω is the set of all possible combination of the variables in X.

The local property in MN is defined in terms of neighbourhood relationship between variables as

$$p(x_i|x - \{x_i\}) = p(x_i|N_i) \tag{3.3}$$

This is also known as *Markovianity* [3][14]. It states that the conditional probability of a node X_i given the rest of the variables can be completely defined in terms of the conditional probability of the node given its neighbouring states N_i. N_i is sometimes referred to as *Markov Blanket* for X_i [19]. As can be noted, unlike BN, there is no obvious ways to obtain a joint distribution from the local conditional probabilities in MN.

3.3 Markov Network Based EDAs

The global Markov property of MN (3.2) has been exploited in most of the MN based EDAs. The underlying concept is to factorise the joint probability distribution in terms of the cliques in the undirected graph and sample it to generate new solutions. Three main categories can be distinguished in this class of Markov network based EDAs. They are:

1. Distribution Estimation using Markov network algorithm (DEUM)
2. Markov Network Estimation of Distribution Algorithm (MN-EDA), Markov network Factorised Distribution Algorithm (MN-FDA)
3. Factorised distribution algorithm (FDA)

Subsequent chapters of this book gives detail on these algorithms.

3.4 Markovianity based Optimisation Algorithm (MOA)

Here we describe an EDA using the local Markov property, the Markovianity. We call it Markovianity based Optimisation Algorithm (MOA). Since, it only exploits the local Markov property, MOA can be seen as the subset of the other global Markov property based EDAs, with a simpler workflow. Furthermore, in addition to gain in efficiency, it avoids the numerical operations associated to the computation of potentials or Kikuchi approximation, which may also represent gains in model accuracy.

Markovianity based Optimisation Algorithm

1. Generate initial (parent) population P of size M
2. Select set D from P consisting of N solutions, where $N <= M$
3. Estimate a Markov network structure from D
4. Estimate local Markov conditional probabilities, $p(x_i|N_i)$, for each variable X_i as defined by the undirected structure and sample them to generate new population
5. Replace old population by new one and go to step 2 until termination criteria are meet

---

Fig. 3.1 The workflow of Markovianity based Optimisation Algorithm

Figure 3.1 shows the workflow of MOA. It starts by generating a population of solutions. A set of solutions is then selected from the population using a selection method, which are then used to estimate the structure of the Markov network. The conditional probabilities defined by the local Markov property (3.3) are then estimated from the selected set of solutions and sampled to generate the new population.

Estimating structure - Step 3 of MOA

1. Create a matrix of mutual information, *MI*, by estimating mutual information for each pair of variables in the solution. Mutual information between two random variables, A and B, is given by

$$CE(A,B) = \sum_{a,b} p(a,b) log\left(\frac{p(a,b)}{p(a)\cdot p(b)}\right)$$

where sum is over all possible combinations of A and B, and $p(a,b)$ is the joint probability of $A = a$ and $B = b$ computed from D

2. Create an edge between two variables, if the mutual information between them is higher than the given threshold. Here we compute the threshold, TR as $TR = avg(MI) * sig$, where $avg(MI)$ is the average of the elements of the MI matrix and sig is the significance parameter, which for the purpose of this paper is set to 1.5.
3. If the number of neighbours to a variable is higher than the maximum number, *MN*, allowed, only keep *MN* neighbours that have the highest mutual information.

Fig. 3.2 The workflow of an undirected structure learning algorithm

A number of different approaches can be used in order to estimate an undirected structure. An entropy based approach is adopted in MOA. More precisely, the mutual information[1] of each pair of variables in the solution is estimated to create a matrix of mutual information. The pairs with the mutual information higher than a certain threshold are then made neighbours. Also, in order to avoid an overly complex network, the number of neighbours that a variable can have is limited to a certain number. Figure 3.2 describes the implemented Markov network structure learning algorithm.

After estimating the structure of the network, the next step is to estimate the conditional probabilities and sample a new population from it. By its definition, an undirected structure may contain cycles. Apart from some restricted set of undirected structures, for example those that satisfy running intersection properties and can be formulated as a directed acyclic graph, most of the Markov networks do not satisfy the ancestral ordering of variables needed by PLS. Alternatively, Markov Chain Monte Carlo (MCMC) [15] methods could be used for sampling. Gibbs sampler [6], a class of MCMC method, is used as the sampling method in MOA. A number of different versions of Gibbs sampler can be implemented for this purpose. Figure 3.3 describes a version that has been implemented in MOA. It starts by generating a random solution. Then for a fixed number of iteration, r, it randomly chooses a variable in the solution, calculates the conditional probability of that variable given the configuration of its neighbouring variables as encoded by the MN structure, and samples it to get the new value for that variable. At the end of the iteration r, the

[1] Mutual information was originally used by [2] to learn tree-based factorisations in EDAs. It has also been used in DEUM to find the MN structure [30].

resulting solution is taken as the new solution. Note that each execution of the Gibbs sampler creates a single solution. Multiple execution of Gibbs sampler should be done in order to create the population of solutions. Also, note that the number of iteration, r is set as the fraction of $nln(n)$, since $nln(n)$ is the approximation of random iteration needed to hit each variable in the solution of size n.

It is also important to notice here that, the conditional probability in general is estimated as

$$p(x_i|N_i) = \frac{p(x_i,N_i)}{p(N_i)} = \frac{p(x_i,N_i)}{\sum_{x_i'} p(x_i',N_i)} \quad (3.4)$$

Assuming variables x_i are binary, this can be written as

$$p(x_i = 1|N_i) = \frac{p(x_i = 1,N_i)}{p(x_i = 1,N_i) + p(x_i = 0,N_i)} \quad (3.5)$$

For the purpose of the temperature based Gibbs sampler in MOA, here the conditional probabilities are estimated as

$$p(x_i|N_i) = \frac{e^{p(x_i,N_i)/T}}{\sum_{x_i'} e^{p(x_i',N_i)/T}} \quad (3.6)$$

Gibbs Sampler - Step 4 of MOA

1. Generate a solution $x = \{x_1, x_2, .., x_n\}$ at random.
2. For r iterations (in this paper we set $r = n \times ln(n) \times IT$, where IT, the *iteration coefficient*, is set to 4), do the following:

   a. Choose a variable x_i from x at random.
   b. Using selected set of solutions, D, compute conditional probabilities $p(x_i|N_i)$ for each value of x_i as Gibbs probability,

   $$p(x_i|N_i) = \frac{e^{p(x_i,N_i)/T}}{\sum_{x_i'} e^{p(x_i',N_i)/T}}$$

   where sum is over all possible values of x_i. For example, in binary case, where $x_i = \{0,1\}$, the probability of $x_i = 1$ given the value of its neighbours N_i is written as

   $$p(1|N_i) = \frac{e^{p(1,N_i)/T}}{e^{p(1,N_i)/T} + e^{p(0,N_i)/T}}$$

   Here, T is the *temperature coefficient* that controls the convergence of the Gibbs probability distribution. Increasing T makes the distribution close to being uniform, and decreasing T converges it to an extremum.
   c. Sample $p(x_i|N_i)$ to get new x_i.

3. Terminate with answer x.

Fig. 3.3 The workflow of implemented Gibbs Sampler algorithm

For binary variables, this can be written as

$$p(1|N_i) = \frac{e^{p(1,N_i)/T}}{e^{p(1,N_i)/T} + e^{p(0,N_i)/T}} \tag{3.7}$$

The benefit of formulating the probabilities in this form is that the temperature coefficient now controls the convergence of the distribution, and therefore can be used to balance the exploration and exploitation of the search space.

In MOA, a linear schedule for the temperature is used. It is defined as $T = \frac{1}{g \times CR}$, where g is the current generation of MOA and CR is the *cooling rate* parameter. CR can be varied in order to control the convergence of MOA. For instance, setting CR higher will result in quick convergence of the conditional probabilities and therefore a quick convergence to a solution, i.e. gain in efficiency, but with less exploration of the search space. Conversely, smaller CR would result in slower convergence of the conditional probabilities and therefore slower convergence to a solution, i.e. more exploration of the search space, but with higher fitness evaluation.

3.5 Results and Analysis

Several key aspects of model building and sampling in MOA have been tested and reported in [35][36][37]. Here we review some of the interesting results. The experiments were performed with several versions of the widely used benchmark deceptive test functions [17][24][31]. They are, deceptive function of order 3 (deceptive3) [7] and the trap function [23]. The *deceptive*3 function is defined as

$$deceptive3(x) = \sum_{i=1}^{\frac{n}{3}} f_{Gdec}(x_{3i-2} + x_{3i-1} + x_{3i}) \tag{3.8}$$

$$f_{Gdec}(u) = \begin{cases} 0.9 \ for \ u = 0 \\ 0.8 \ for \ u = 1 \\ 0.0 \ for \ u = 2 \\ 1.0 \ for \ u = 3 \end{cases}$$

Where, u is the number of ones in the input block of 3 bits.

Similarly, a Trap function of order k can be defined as

$$f_{trap,k}(x) = \sum_{i=1}^{n/k} trap_k(x_{b_i,1} + ... + x_{b_i,k}) \tag{3.9}$$

Each block $(x_{b_i,1} + ... + x_{b_i,k})$ gives a fitness which can be calculated through a general trap function of order k

$$trap_k(u) = \begin{cases} f_{high}, & \text{if} \quad u = k \\ f_{low} - u\frac{f_{low}}{k-1}, & \text{otherwise} \end{cases}$$

Where, u is the number of ones in the input block of k bits, and f_{high} and f_{low} are parameters that control the distance between the local and global optima. .

In order to test the different aspects of MOA performance, the results are divided into five parts.

3.5.1 Comparison with Other EAs

Here the performance of MOA is compared against the performance of its Bayesian network counterpart, the BOA. Comparison is made with the BOA results reported in [24]. BOA has been shown to significantly outperform GA in both of these functions. Therefore, it was also interesting to compare the performance of MOA with GA.

For both deceptive3 and trap function, problem size, n, ranged from 30 to 360 bits. The order of interaction in trap function was set to 5, (trap5). It is an instance of the general trap function, where $k = 5$, $f_{high} = 5$ and $f_{low} = 4$. For all experiments, population size (PS) was gradually increased until all of the 10 runs of the algorithm found the optimal solution.

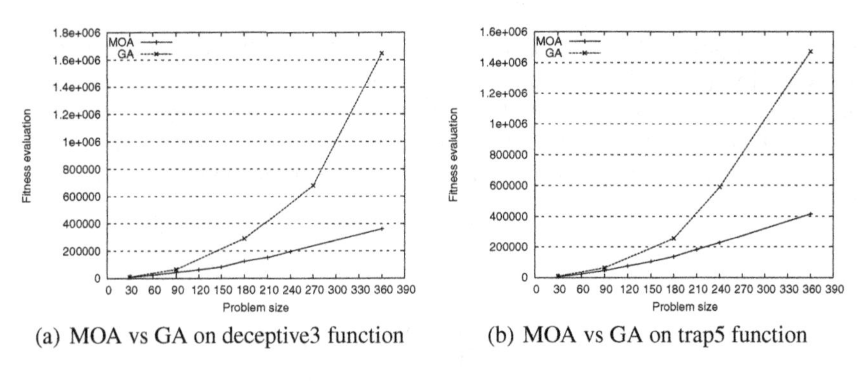

(a) MOA vs GA on deceptive3 function (b) MOA vs GA on trap5 function

Fig. 3.4 Scalability graph comparing the performance of MOA and GA for both deceptive and trap function of size ranging from 30 bits to 360 bits

Figure 4(a) shows the scalability graph comparing the performance of both MOA and GA over different problem sizes for deceptive problem and Figure 4(b) shows the same for trap problem. The results showed that, for smaller problems, the performance of MOA and GA was comparable, however once the problem size starts to get larger, MOA significantly outperformed GA in terms of number of fitness evaluations required to find the optimal solution. Also, the lower standard deviation for the fitness evaluation in MOA suggested that, it was a more predictable algorithm than GA.

Also, in comparison to BOA, MOA required slightly less fitness evaluations. This could be observed by comparing the MOA results with that presented in [24] for BOA. As an example, it was shown in [24] that BOA requires in average around 160000 fitness evaluations to solve 180 bit deceptive function, while MOA only requires around 125000 fitness evaluations. Also, for trap function, BOA in average required around 220000 fitness evaluations, while MOA only required around 136000 fitness evaluations[2].

3.5.2 Introducing Permutation

The aim here is to show that different ordering of the bits in the solution does not affect the performance of the MOA. For this purpose, the ordering of the bits in the solution for Trap5 problem was randomly permuted. An example of original and modified dependency graph is shown in figure 3.5 (b) and figure 3.5 (c) respectively. This makes the bits in a single block further apart from each other. This obviously has a negative effect to the performance of the GA, since the crossover operator does not take into account the dependencies between the variables in the solution and can easily disrupt a correct configuration of the block. Both GA and MOA were tested with the permuted trap5 function. The parameter setups were the same as in previous section. As expected, GA was not able to find the solution, even for the very small problem size of 30 bits and with a very high population size of 20000. However, MOA as expected was able to find the solution in similar fitness evaluation as with non-permuted trap5, as shown in figure 4(b).

3.5.3 Introducing Overlaps

Next, the overlap between the blocks in the solution is introduced. For example, overlap of order one means a variable in a block is common to the next block. Figure 3.5 (d) shows an example of the overlap of order one in trap function of order 4 (trap4). Notice that there is a cycle, i.e., the last block overlaps with the first block. This significantly increases the complexity of already difficult to solve deceptive problems. Here, not only finding the correct configuration of values in block is difficult, but once found, preserving them is also very difficult, since it can be easily disrupted by the incorrect configuration of the neighbouring blocks. Note that, with the introduction of overlaps, the number of blocks also increases in comparison to non overlapping problem of the same size. For example, there are 15 blocks of order 4 in 60 bit non-overlapping trap4 function. In contrast, there are 20 blocks of order 4 in the order one overlapping trap4 problem.

MOA was tested on order one overlapping trap problems with $k = 4$ (trap4). The problem size ranged from 30 to 120 bits. As with previous case, GA was not able to find the solution even for the very small sized problem of 30 bits. The parameter setups for MOA were as follows: population size (PS) ranged from 600 to 6000 for 30 to 120 bit problems. Truncation selection with selection size (SS) of 50% of the

[2] Note that in order to make the comparison fair, the performance of MOA was compared with the version of BOA [24], which also had a parameter similar to MN that restricted the maximum number of parents going to a node. Similar to the later version of BOA, improved structure learning algorithm is likely to remove this parameter from MOA workflow.

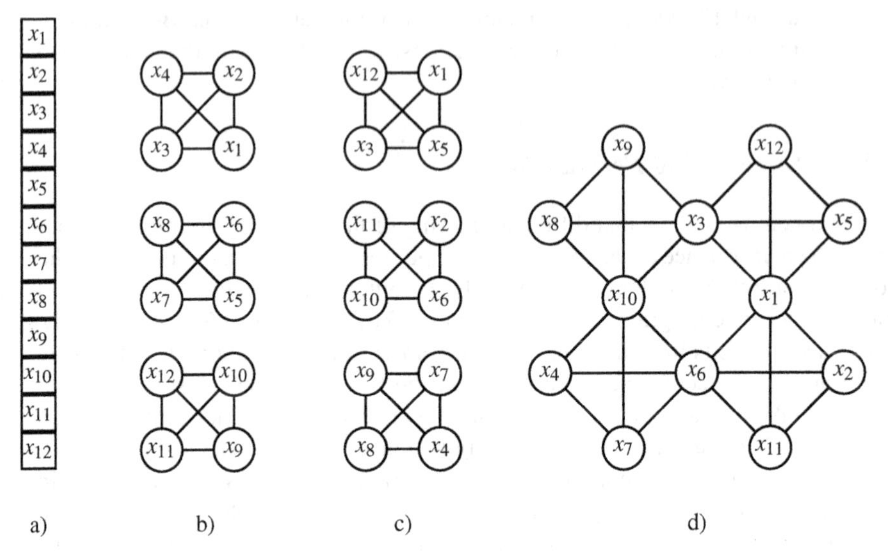

Fig. 3.5 An order 4 deceptive problem with different dependency structure. a) The ordering of the variables in the solution. b) The original dependency. c) A permuted dependency. d) An overlapping permuted dependency.

PS was used. In order to prevent quick diversity loss, the elitism parameter (*EL*) was set to the 50% of the *PS*, i.e., best half of the parent population was preserved in the next generation. Also, the number of maximum neighbours (*MN*) allowed to the structure of the Markov network in MOA was set to 6, since (as can be seen from the Figure 3.5 (d)) there can be at most 6 neighbours for a variable.

The scalability graph showing the number of fitness evaluations required by MOA to find the solution for 30 to 120 bit permuted trap4 problem with overlap of order 1 is shown in Figure 3.6. The same statistic for the MOA on non-permuted trap4 function with overlap of order 1 is also plotted in order to (again) show that ordering of bits in the solution does not make difference to the performance of MOA[3].

3.5.4 *Incorporating Prior Information*

Prior information about the problem can be a valuable input in improving the performance of the algorithm. This is particularly relevant with EDAs, since they explicitly model the interaction between variables. Many real world problems exhibit bidirectional interaction between variables. This section shows how the prior information about this natural dependency can be incorporated in MOA, and also analyses its

[3] Note that in [25], BOA has been tested on a multivariate function called random decomposable problems (rADPs). While, this function also has overlapping dependency, it does not consider deceptiveness and therefore has different properties.

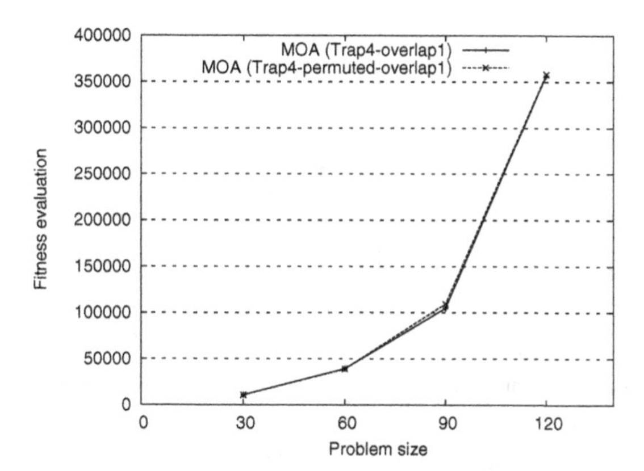

Fig. 3.6 Number of fitness evaluations required by MOA on overlapping trap4 function of order 1 for size ranging from 30 to 120 bits with both permuted ordering and non-permuted ordering of bits

effect to the performance of the algorithm. For this, MOA was tested on different deceptive functions shown in Figure 3.5, by giving their dependency as an undirected graph. We note that with a Bayesian network based EDA, finding exact DAG structure for deceptive problems (in particular, to the one with cycles, Figure 3.5 (d)) can be difficult [4]. While with MOA, such dependency can be directly represented as a Markov network. Results comparing the performance of MOA with pre-given structure and without pregiven structure is shown in figure 7(a) for deceptive3 function, in figure 7(b) for trap5 function and in figure 7(c) for overlapping trap4 function.

It can be seen that, for deceptive3 problem, there was no improvement in the performance, while for trap5 problem the performance improvement was significant. Again, for overlapping trap4 problem the improvement in performance was marginal. These results suggest that, giving the prior information about the dependency may improve the performance of the algorithm. However, this is not guaranteed and is highly problem-dependent. These results are interesting and clearly require further work in order to get the explanation of these effects. These results also link us back to an open question in the EDA that is whether it is necessary to have an exact problem structure in order to get a better performance?

3.5.5 Gibbs Sampling vs. Temperature Less Sampling

Gibbs sampling with temperature coefficient is integral part of MOA workflow that distinguishes it from other local property based MN EDAs. This section compares the performance of MOA with Gibbs sampling with MOA with *temperature less sampling*. The workflow of temperature less sampling is the same as the one shown in Figure 3.3, except for the conditional probability estimation part, which is done as

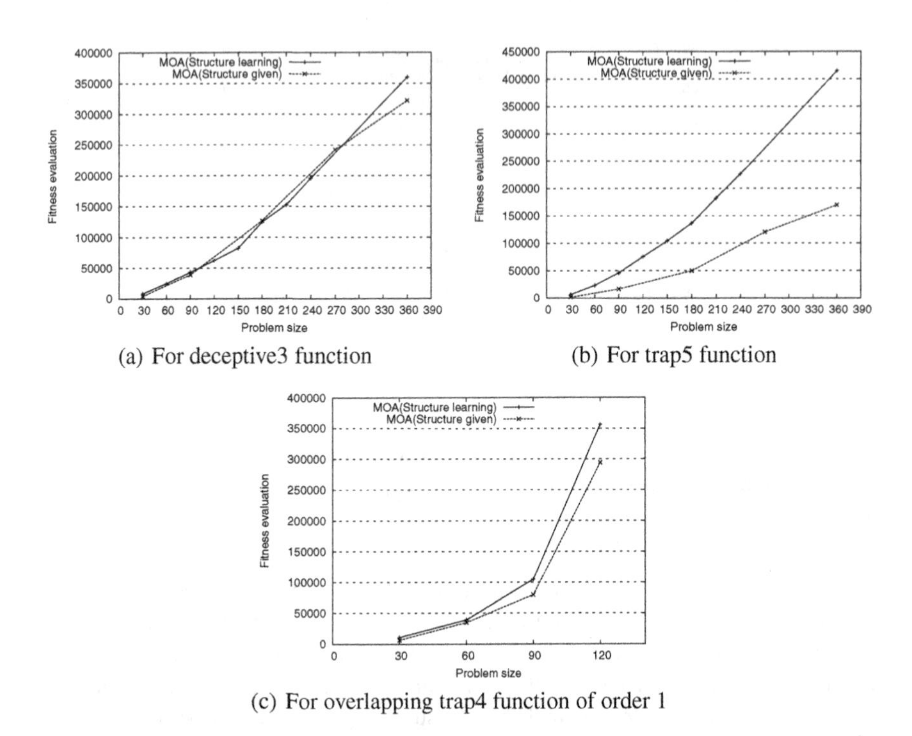

(a) For deceptive3 function

(b) For trap5 function

(c) For overlapping trap4 function of order 1

Fig. 3.7 Scalability graph showing the number of fitness evaluations required by MOA when the structure of the problem is learnt from the population and when the structure of the problem is given as the prior information

$$p(x_i|N_i) = \frac{p(x_i,N_i)}{\sum_{x_i'} p(x_i',N_i)} = \frac{p(x_i,N_i)}{p(N_i)}$$

It was found that temperature less sampling was not able to find the solution for higher sized problems within a reasonable population size. Figures 8(a) show a typical example of how maximum and average fitness in the population progress in each generation during a typical run of MOA with Gibbs sampling for 60 bit trap function. Similarly, Figure 8(b) shows the same for the temperature less sampling.

It can be noticed that the temperature less sampling could not find the optimal solution in 50 generations and converged to some near optimal solution, where as Gibbs sampling finds the solution in around 30 generations. We can also see that the curves in Gibbs sampling is of sigmoid shape in comparison to (near) linear shape in temperature less sampling. This suggests that temperature less sampling narrows the search space as generation progress. In contrast, Gibbs sampling first (thoroughly) explores the search space and then converges to a solution. Cooling rate (CR) parameter influences the form of the curve in the Gibbs sampling. The lower the CR, the higher the exploration and more sigmoid the curve is.

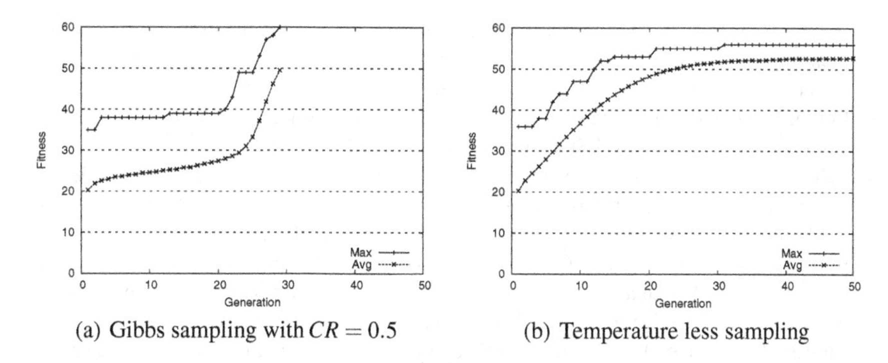

(a) Gibbs sampling with $CR = 0.5$ (b) Temperature less sampling

Fig. 3.8 An example of how maximum and average fitness in the population progress in each generation during a typical run of MOA on 60 bit trap function

3.6 Conclusion

Interaction between variables in many real world problems can be naturally represented as an undirected graph. Well known examples of such problems include Ising spin glasses and SAT. Structure of such problems can be readily incorporated in MOA without requiring to have a structure learning step. This is an important property of MOA that distinguishes it from other Bayesian network based EDAs.

As stated earlier, it is important to also distinguish the difference between MOA and its other Markov network based counterparts, DEUM and MN-EDA. In particular, DEUM defines clique potential functions to encapsulate interaction between variable and builds a model of fitness function to approximate the joint probability distribution of a Markov network. Similarly, MN-EDA also estimates joint probability distribution by means of Kikuchi approximation. In contrast, MOA estimates conditional probabilities defined by the neighbourhood structure of the Markov network and does not estimates joint probability distribution. These conditional probabilities are then sampled using Gibbs sampler. A range of reviewed experiments that tests the different aspects of MOA performance confirms that MOA can effectively solve problems with complex interaction between variables.

Acknowledgements. This work has been partially supported by the Saiotek and Research Groups 2007-2012 (IT-242-07) programs (Basque Government), TIN2010-14931 and Consolider Ingenio 2010 - CSD 2007 - 00018 projects (Spanish Ministry of Science and Innovation).

References

1. Alden, M.A.: MARLEDA: Effective Distribution Estimation Through Markov Random Fields. PhD thesis, Faculty of the Graduate Schoool, University of Texas at Austin, USA (December 2007)

2. Baluja, S., Davies, S.: Using optimal dependency-trees for combinatorial optimization: Learning the structure of the search space. In: Proceedings of the 14th International Conference on Machine Learning, pp. 30–38. Morgan Kaufmann (1997)
3. Besag, J.: Spatial interactions and the statistical analysis of lattice systems (with discussions). Journal of the Royal Statistical Society 36, 192–236 (1974)
4. Echegoyen, C., Lozano, J.A., Santana, R., Larrañaga, P.: Exact Bayesian network learning in estimation of distribution algorithms. In: Proceedings of the 2007 Congress on Evolutionary Computation CEC 2007, pp. 1051–1058. IEEE Press (2007)
5. Etxeberria, R., Larrañaga, P.: Global optimization using Bayesian networks. In: Ochoa, A., Soto, M.R., Santana, R. (eds.) Proceedings of the Second Symposium on Artificial Intelligence (CIMAF 1999), Havana, Cuba, pp. 151–173 (1999)
6. Geman, S., Geman, D.: Stochastic relaxation, Gibbs distributions and the Bayesian restoration of images. In: Fischler, M.A., Firschein, O. (eds.) Readings in Computer Vision: Issues, Problems, Principles, and Paradigms, pp. 564–584. Kaufmann, Los Altos (1987)
7. Goldberg, D.: Genetic Algorithms in Search, Optimization, and Machine Learning. Addison-Wesley (1989)
8. Henrion, M.: Propagating uncertainty in Bayesian networks by probabilistic logic sampling. In: Lemmer, J.F., Kanal, L.N. (eds.) Uncertainty in Artificial Intelligence 2, pp. 149–163. North-Holland, Amsterdam (1988)
9. Kikuchi, R.: A Theory of Cooperative Phenomena. Physical Review 81, 988–1003 (1951)
10. Larrañaga, P., Etxeberria, R., Lozano, J.A., Peña, J.M.: Combinatorial optimization by learning and simulation of Bayesian networks. In: Proceedings of the Sixteenth Conference on Uncertainty in Artificial Intelligence, Stanford, pp. 343–352 (2000)
11. Larrañaga, P., Lozano, J.A.: Estimation of Distribution Algorithms: A New Tool for Evolutionary Computation. Kluwer Academic Publishers (2002)
12. Lauritzen, S.L.: Graphical Models. Oxford University Press (1996)
13. Lauritzen, S.L., Spiegelhalter, D.J.: Local computations with probabilities on graphical structures and their application to expert systems. Journal of the Royal Statistical Society B 50, 157–224 (1988)
14. Li, S.Z.: Markov Random Field modeling in computer vision. Springer (1995)
15. Metropolis, N.: Equations of state calculations by fast computational machine. Journal of Chemical Physics 21, 1087–1091 (1953)
16. Mitchell, M.: An Introduction To Genetic Algorithms. MIT Press, Cambridge (1997)
17. Mühlenbein, H., Mahnig, T., Ochoa, A.R.: Schemata, distributions and graphical models in evolutionary optimization. Journal of Heuristics 5(2), 215–247 (1999)
18. Mühlenbein, H., Paaß, G.: From Recombination of Genes to the Estimation of Distributions: I. Binary Parameters. In: Ebeling, W., Rechenberg, I., Voigt, H.-M., Schwefel, H.-P. (eds.) PPSN 1996. LNCS, vol. 1141, pp. 178–187. Springer, Heidelberg (1996)
19. Murphy, K.: Dynamic Bayesian Networks: Representation, Inference and Learning. PhD thesis, University of California, Berkeley (2002)
20. Murray, I., Ghahramani, Z.: Bayesian Learning in Undirected Graphical Models: Approximate MCMC algorithms. In: Twentieth Conference on Uncertainty in Artificial Intelligence (UAI 2004), Banff, Canada, July 8-11 (2004)
21. Ochoa, A., Soto, M.R., Santana, R., Madera, J., Jorge, N.: The factorized distribution algorithm and the junction tree: A learning perspective. In: Ochoa, A., Soto, M.R., Santana, R. (eds.) Proceedings of the Second Symposium on Artificial Intelligence (CIMAF 1999), Havana, Cuba, March 1999, pp. 368–377 (1999)

22. Pearl, J.: Probabilistic Reasoning in Intelligent Systems. Morgan Kaufman Publishers, Palo Alto (1988)
23. Pelikan, M.: Bayesian optimization algorithm: From single level to hierarchy. PhD thesis, University of Illinois at Urbana-Champaign, Urbana, IL, Also IlliGAL Report No. 2002023 (2002)
24. Pelikan, M., Goldberg, D.E., Cantú–Paz, E.: BOA: The Bayesian Optimization Algorithm. In: Banzhaf, W., et al. (eds.) Proceedings of the Genetic and Evolutionary Computation Conference GECCO 1999, vol. I, pp. 525–532. Morgan Kaufmann, San Francisco (1999)
25. Pelikan, M., Sastry, K., Butz, M.V., Goldberg, D.E.: Hierarchical BOA on random decomposable problems. IlliGAL Report No. 2006002, University of Illinois at Urbana-Champaign, Illinois Genetic Algorithms Laboratory, Urbana, IL (January 2006)
26. Santana, R.: A Markov Network Based Factorized Distribution Algorithm for Optimization. In: Lavrač, N., Gamberger, D., Todorovski, L., Blockeel, H. (eds.) ECML 2003. LNCS (LNAI), vol. 2837, pp. 337–348. Springer, Heidelberg (2003)
27. Santana, R.: Estimation of Distribution Algorithms with Kikuchi Approximations. Evolutionary Computation 13, 67–98 (2005)
28. Santana, R., Bielza, C., Larrañaga, P., Lozano, J.A., Echegoyen, C., Mendiburu, A., Armanñanzas, R., Shakya, S.: MATEDA 2.0: Estimation of distribution algorithms in MATLAB. Journal of Statistical Software 35(7), 1–30 (2010)
29. Shakya, S.: DEUM: A Framework for an Estimation of Distribution Algorithm based on Markov Random Fields. PhD thesis, The Robert Gordon University, Aberdeen, UK (April 2006)
30. Shakya, S., Brownlee, A., McCall, J., Fournier, F., Owusu, G.: DEUM – A Fully Multivariate EDA Based on Markov Networks. In: Chen, Y.-p. (ed.) Exploitation of Linkage Learning. ALO, vol. 3, pp. 71–93. Springer, Heidelberg (2010)
31. Shakya, S., McCall, J.: Optimisation by Estimation of Distribution with DEUM framework based on Markov Random Fields. International Journal of Automation and Computing 4, 262–272 (2007)
32. Shakya, S., McCall, J., Brown, D.: Updating the probability vector using MRF technique for a univariate EDA. In: Onaindia, E., Staab, S. (eds.) Proceedings of the Second Starting AI Researchers' Symposium. Frontiers in Artificial Intelligence and Applications, vol. 109, pp. 15–25. IOS press, Valencia (2004)
33. Shakya, S., McCall, J., Brown, D.: Using a Markov Network Model in a Univariate EDA: An Emperical Cost-Benefit Analysis. In: Proceedings of Genetic and Evolutionary Computation COnference (GECCO 2005), pp. 727–734. ACM, Washington, D.C. (2005)
34. Shakya, S., McCall, J., Brown, D.: Solving the Ising spin glass problem using a bivariate EDA based on Markov Random Fields. In: Proceedings of IEEE Congress on Evolutionary Computation (IEEE CEC 2006), pp. 3250–3257. IEEE Press, Vancouver (2006)
35. Shakya, S., Santana, R.: An EDA based on local Markov property and Gibbs sampling. In: Proceedings of Genetic and Evolutionary Computation COnference (GECCO 2008). ACM Press, Atlanta (2008)
36. Shakya, S., Santana, R.: A markovianity based optimisation algorithm. Technical Report Technical Report EHU-KZAA-IK-3/08, Department of Computer Science and Artificial Intelligence, University of the Basque Country (September 2008)
37. Shakya, S., Santana, R.: A markovianity based optimisation algorithm. Genetic Programming and Evolvable Machines (2011) (in press)

Chapter 4
DEUM - Distribution Estimation Using Markov Networks

Siddhartha Shakya, John McCall, Alexander Brownlee, and Gilbert Owusu

Abstract. DEUM is one of the early EDAs to use Markov Networks as its model of probability distribution. It uses undirected graph to represent variable interaction in the solution, and builds a model of fitness function from it. The model is then fitted to the set of solutions to estimate the Markov network parameters; these are then sampled to generate new solutions. Over the years, many different DEUM algorithms have been proposed. They range from univariate version that does not assume any interaction between variables, to fully multivariate version that can automatically find structure and build fitness models. This chapter serves as an introductory text on DEUM algorithm. It describes the motivation and the key concepts behind these algorithms. It also provides workflow of some of the key DEUM algorithms.

4.1 Introduction

Probabilistic graphical models are a useful tool in EDA for modelling and sampling probability distributions. Particularly, directed graphical models (Bayesian networks) [23] have been widely used in EDAs. Some of the well known instances of Bayesian network based EDA includes Bayesian Optimisation Algorithm (BOA) [24], hierarchical Bayesian Optimisation Algorithm (hBOA) [25], Estimation of Bayesian Network Algorithm (EBNA) [8][12] and Learning Factorised Dis-

Siddhartha Shakya · Gilbert Owusu
Business Modelling and Operational Transformation Practice,
BT Innovate & Design, Ipswich, UK
e-mail: sid.shakya@bt.com, gilbert.owusu@bt.com

Alexander Brownlee
Loughborough University, Loughborough, UK
e-mail: a.e.i.brownlee@lboro.ac.uk

John McCall
School of Computing, Robert Gordon University, Aberdeen, UK
e-mail: jm@comp.rgu.ac.uk

S. Shakya and R. Santana (Eds.): Markov Networks in Evolutionary Computation, ALO 14, pp. 55–71.
springerlink.com

tribution Algorithm (LFDA) [18]. Recent years have seen an increasing interest in the use of undirected graphical models (Markov networks) [2][13][20] in EDAs [27][34][36][28][31][29][33][6][7]. Some of the well known instances of Markov network based EDA includes Distribution Estimation Using Markov Networks (DEUM) algorithm [31], Markov Network EDA (MN-EDA) [28], Markov Network Factorised Distribution Algorithm (MN-FDA) [27] and Markovianity based Optimisation Algorithm (MOA) [40–42].

This chapter give introduction to DEUM algorithms [34][31]. They are a family of Markov network based EDA that builds a model of fitness function from the undirected graphs, and use this model to estimate the parameters of the Gibbs distribution. Markov chain Monte Carlo simulations, including Gibbs sampler [9] and Metropolis sampler [17], are then used to sample new solutions from the Gibbs distribution. Several variants of DEUM have been proposed and are found to perform well in comparison to other EDAs of their class in a range of different test problems including Ising Spin Glass and SAT. [34][36][37][5]. These algorithms can be categorised into two key classes, a) fixed structure and b) structure learning. As name suggests, fixed structure DEUM algorithms assume the interaction between variables in the problem to be pre-given. In other words, they do not learn the structure of the problem and assume that it is known in advance. Structure learning DEUM can automatically learn the undirected structure of the problem, automatically find the cliques from the structure and automatically estimate the joint probability model of the Markov network. In this chapter we review both types of DEUM algorithms. This chapter is intended as an introductory text on DEUM algorithms. Some theoretical aspects of fitness modelling in DEUM is described Chapter 8 and the applications of DEUM is described in Chapter 8.

The outline of the chapter is as follows. Section 2 briefly reviews Markov network and its key properties. Section 3 describes the fitness modelling approach to estimating the parameters of the Markov network, a key property of DEUM algorithm that distinguishes it from other Markov network based EDAs. It also describes sampling techniques used by DEUM algorithms. Section 4 describes DEUM$_d$ algorithm, a univariate DEUM algorithm that samples from Gibbs distribution. Section 5 describes Is-DEUM, a multivariate DEUM with a fixed grid structure. Section 6 describes the structure learning full DEUM algorithm that can automatically learn the structure, build fitness models and sample from it. Section 7 concludes the chapter.

4.2 Markov Networks

A Markov network (also known as Markov Random Fields or Undirected Graphical models) regards a solution, $x = \{x_1, x_2, .., x_n\}$, as a set of values taken by a set of random variables, $X = \{X_1, X_2, ..., X_n\}$. The network itself is a pair (G, Ψ), where G is the structure and the Ψ is the parameter set of the network. G is an undirected graph where each node corresponds to a random variable X_i in the modelled data set X, and each edge corresponds to a conditional dependency between two variables. However, unlike Bayesian networks, the edges in Markov networks are undirected. Here, the relationship between two nodes should be seen as a *neighbourhood relationship*, rather than a *parenthood relationship* in Bayesian Networks. We use $N = \{N_1, N_2, ..., N_n\}$ to define a *neighbourhood system* on G, where each N_i is a set of nodes neighbouring to node X_i. Figure 4.1 shows an example of a Markov network structure on 6 random variables. Here, variable X_1 has 2 neighbours, $N_1 = \{X_2, X_3\}$. Similarly, variable X_2 has 4 neighbours $N_2 = \{X_1, X_3, X_4, X_5\}$.

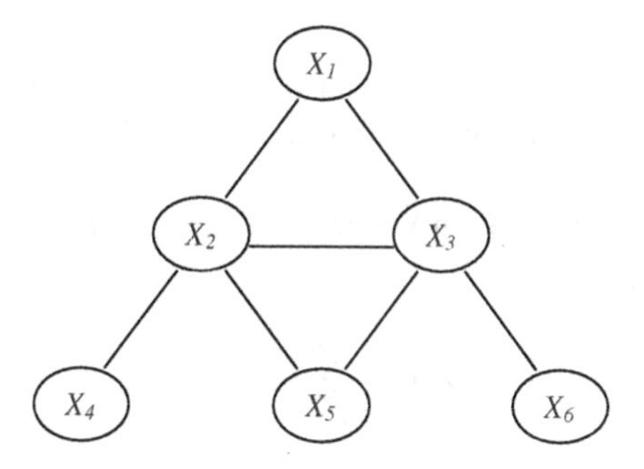

Fig. 4.1 A Markov network structure on 6 random variables

Let us once again review local and global properties of Markov network[1]. A Markov network is characterised in terms of neighbourhood relationship between variables by its *local Markov property* known as *Markovianity* [2][13], which states that the conditional probability of a node X_i given the rest of the variables can be completely defined in terms of it's conditional probability given only its neighboring states N_i. N_i is sometimes referred to as *Markov Blanket* for X_i [19]. In terms of probability, it can be written as

$$p(x_i|x - \{x_i\}) = p(x_i|N_i) \tag{4.1}$$

A Markov network is also characterised in terms of *cliques*[2] in the undirected graph by its global property, the joint probability distribution, and can be written as

$$p(x) = \frac{1}{Z}\prod_{i=1}^{m}\psi_i(c_i) \tag{4.2}$$

where, $\psi_i(c_i)$ (or more precisely $\psi_i(C_i = c_i)$) is a positive *potential function* on clique $C_i \in X$, m is the number of cliques in the structure G. $Z = \sum_{x\in\Omega}\prod_{i=1}^{m}\psi_i(c_i)$ is the normalising constant known as the *partition function* which ensures that $\sum_{x\in\Omega} p(x) = 1$. Here, Ω is the set of all possible combination of the variables in X.

[1] See Chapter 1 for more detail on Markov Networks.

[2] Given an undirected graph G, a clique is a fully connected subset of nodes. For example, in Figure 4.1, variables $\{X_1, X_2, X_3\}$ define a clique.

Equivalently, using Hammersley-Clifford theorem [10], the global Markov property can also be written in terms of Gibbs distribution as

$$p(x) = \frac{e^{-U(x)/T}}{Z} \tag{4.3}$$

where,

$$Z = \sum_{y \in \Omega} e^{-U(y)/T} \tag{4.4}$$

is a normalising constant, T is a parameter of the Gibbs distribution known as the *temperature* and $U(x)$ (or more precisely $U(X = x)$) is known as the *energy* of the distribution.

Given an undirected graph, G, on X, energy, $U(x)$, is defined as a sum of *potential functions* over the cliques, C_i, in G.

$$U(x) = \sum_{i=1}^{m} u_i(c_i) \tag{4.5}$$

Here, $u_i(c_i)$ (or more precisely $u_i(C_i = c_i)$) is a potential function defined over a clique $C_i \in X$. Equation (4.3), in terms of clique potential function, can also be written as

$$p(x) = \frac{e^{-\sum_{i=1}^{m} u_i(c_i)/T}}{Z} \tag{4.6}$$

Note that the relationship between $\psi_i(c_i)$ in (4.2) and $u_i(c_i)$ in (4.6) is defined as

$$\psi_i(c_i) = e^{-u_i(c_i)/T} \tag{4.7}$$

The clique potential function, $u_i(c_i)$, captures the way variables interact in the clique c_i. It should be carefully defined in order to get a desired behaviour from a Markov network. This is exactly what DEUM algorithms does, which is described next.

4.3 Fitness Modelling and DEUM Algorithms

As with other EDAs, a general DEUM algorithm starts by initialising a population of parent solutions, P. It then selects a set of promising solution D from P, which is then used to estimate a Markov network. The estimated network is then sampled to generate new solutions. These new solutions replace the parent solutions and this iteration continues. While estimating a Markov network, DEUM builds a model of fitness function from the undirected relationship captured by the Markov network structure and fits the model to the selected set of solution to estimate the model parameters. These parameters fully specify the joint probability of the Markov network. This fitness modelling approach to parameterise a Markov network is distinguishing feature of DEUM algorithm, which we describe next.

4.3.1 Fitness Modelling

Assuming that the probability of a solution is proportional to its fitness, the jpd, $p(x)$, can be modelled in terms of fitness as

$$p(x) = \frac{f(x)}{Z} \tag{4.8}$$

where, $f(x) > 0$, and $Z = \sum_{y \in \Omega} f(y)$ is the partition function and Ω is the set of all possible solutions.

Now, from (4.3) and (4.8), we can deduce following equivalence of jpd for Markov networks in terms of fitness function.

$$p(x) = \frac{e^{-U(x)/T}}{\sum_{y \in \Omega} e^{-U(y)/T}} \equiv \frac{f(x)}{\sum_{y \in \Omega} f(y)} \tag{4.9}$$

From which, following relationship between fitness and the energy can be deduced [4].

$$-ln(f(x)) = U(x) \tag{4.10}$$

For simplicity, here we assume T from (4.9) to be 1. In other words, (4.10) defines the equivalence shown in (4.9). We refer to (4.10) as **M**arkov network **F**itness **M**odel (**MFM**). From (4.5), MFM can also be written in terms of potential functions as:

$$-ln(f(x)) = \sum_{i=1}^{m} u_i(c_i) \tag{4.11}$$

Energy, $U(x)$, in MFM (4.10) gives the full specification of the jpd (4.3), so MFM can be regarded as a probabilistic model of the fitness function. Also notice that the negative relationship between $U(x)$ and $ln(f(x))$ means that minimising energy $U(x)$ here is equivalent to maximising fitness $f(x)$. Obviously, the negative relationship should be maintained for this purpose, i.e, condition $f(x) > 1$ should be maintained[3]. This is an important property which is used in DEUM algorithm to find the minimum energy configuration that maximise fitness.

At this point, it is important to notice that the log-linear form of MFM (4.11) is the result of our assumption of JPD as a mass distribution of fitness over solution space, as shown in (4.8). We could easily get different relationship between $f(x)$ and $U(x)$ by making different assumption about mass distribution of fitness function. For example, assuming $p(x) = \frac{e^{-f(x)}}{\sum_{y \in \Omega} e^{-f(y)}}$, we would get a linear MFM as $f(x) = \sum_{i=1}^{m} u_i(c_i)$.

In general, the form of energy, $U(x)$ in MFM models the different order of interaction between variables in X.

[3] Notice that if $f(x) < 1$, $ln(f(x))$ will be a negative number, and therefore $-ln(f(x))$ will be a positive number, resulting in negative condition not being met. In DEUM algorithms, $f(x)$ is mapped such that $f(x) > 1$ is maintained.

4.3.2 Univariate MFM

Assuming each variables $X_i \in X$ to be independent, the graph G will be an edge less graph. Therefore, the set of maximal cliques, C, in G would consist of n singleton cliques $C_i = \{X_i\}$. For each clique, $\{X_i\}$, a potential function is associated as follows:

$$u_i(x_i) = \alpha_i x_i \tag{4.12}$$

From (4.10) and (4.11), the univariate MFM can then be written as:

$$-ln(f(x)) = U(x) = \alpha_1 x_1 + \alpha_2 x_2 + ... + \alpha_n x_n \tag{4.13}$$

Here, α_i are the unknown parameters associated with each cliques $\{X_i\}$. They completely specifies $U(x)$ in (4.13), and therefore completely specifies the Gibbs distribution (4.3) as $p(x) = \frac{e^{-\sum_{i=1}^{n} \alpha_i x_i / T}}{Z}$. They are known as Markov network (or Markov Random Field) parameters (MRF parameters) [13]. We use θ to refer to vector of all MRF parameters in the model. For univariate case, the vector $\theta = \alpha = \{\alpha_1, \alpha_2, ..., \alpha_n\}$. In terms of MFM, (4.10), a MRF parameter measures the effect that the interaction between variables in a clique have on the fitness of the solution, $f(x)$. Obviously, in univariate case (4.13), α_i measures the effect of a single variable, X_i, on fitness.

4.3.3 Multivariate MFM

A multivariate MFM models the interaction between two or more variables. Figure 4.2 shows one such structure, where a variable interacts with 4 of its immediate Neighbours. This structure can also be seen as an instance of *Ising model* on two

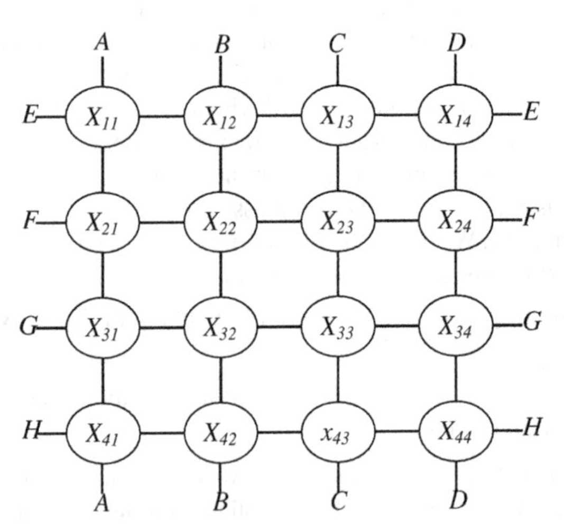

Fig. 4.2 A structure showing the interaction between variables in a two dimensional lattice

dimensional lattice [11]. The set of maximal cliques, C, in this case, contains 2×4^2 bivariate cliques $C_{ij,i'j'} = \{X_{ij}, X_{i'j'}\}$. This structure can be generalised to $n = l \times l$ variables, where C will contain $m = 2l^2$ bivariate cliques. For each clique $\{X_{ij}, X_{i'j'}\}$, a potential function can be assigned as $\beta_{ij,i'j'} x_{ij} x_{i'j'}$, where, each $\beta_{ij,i'j'}$ is the MRF parameter associated with bivariate clique $\{X_{ij}, X_{i'j'}\}$. The energy, $U(x)$ in MFM (4.10) for such X will therefore be

$$-ln(f(x)) = U(x) = \sum_{i=1}^{l} \sum_{j=1}^{l} \left(\beta_{ij,(i+1)j} x_{ij} x_{(i+1)j} + \beta_{ij,i(j+1)} x_{ij} x_{i(j+1)}\right) \tag{4.14}$$

The vector $\theta = \beta$ is used to denote the set of all $2n$ bivariate MRF parameters $\beta_{ij,i'j'}$. In [5] this idea is further extended to include cliques with three variables for solving 3-SAT problems.

4.3.4 Estimating MRF Parameters

Once the MFM is built, next step in DEUM is to estimate the set of all MRF parameters, θ . In general, each solution in a given population provides an equation satisfying the MFM (eg. (4.13), or (4.14)), where θ are the unknows. Selecting a set of solution D consisting of N promising solutions from a population P therefore allows us to estimate these parameters by solving the system of equations[4]:

$$F = A\theta^T \tag{4.15}$$

Here, F is the vector containing $-\ln(f(x))$ of all solutions in D, θ, the unknown part of the equation is the vector of all MRF parameters and A is the matrix of allele values in D. In other words, DEUM fit the MFM to a dataset D and approximate the parameters of the Markov network, θ.

4.3.5 Sampling Markov Networks

Once the MRF parameters are estimated, the joint probability distribution is fully specified. It is then sampled to generate new solution. For this the marginal probability is first derived from the joint distribution as follows.

Let us use x^+ to denote a solution x having a particular $x_i = +1$ and x^- to denote a solution x having $x_i = -1$. The probability that the value of the variable in position i is equal to 1 given its neighbours, $p(x_i = 1|N_i)$, can then be written as

$$p(x_i = 1|N_i) = \frac{p(x^+)}{p(x^+) + p(x^-)} \tag{4.16}$$

[4] In DEUM, Singular Value Decomposition (SVD) [26] is used to solve the system of equations.

Substituting $p(x)$ from (4.3) and cancelling the Z, we get

$$p(x_i = 1|N_i) = \frac{e^{-U(x^+)/T}}{e^{-U(x^+)/T} + e^{-U(x^-)/T}} \tag{4.17}$$

Since, $U(x^+)$ and $U(x^-)$ agree in all terms other than those containing x_i, the common terms in both $U(x^+)$ and $U(x^-)$ drop out and we get much simpler expression as the estimate of the marginal probability for $x_i = 1$ conditional upon N_i:

$$p(x_i = 1|N_i) = \frac{1}{1 + e^{W_i/T}} \tag{4.18}$$

Here, W_i is the difference in two energies, $U(x^+)$ and $U(x^-)$, after substituting the x_i to 1 for all the remaining terms in $U(x^+)$ and to -1 for all remaining terms in $U(x^-)$[5]. For example, W_i for the univariate MFM (4.13) simplifies to

$$W_i = 2\alpha_i \tag{4.19}$$

Similarly, W_i (or more precisely W_{ij}) for multivariate MFM (4.14) simplifies to

$$W_{ij} = \beta_{ij,(i+1)j}x_{(i+1)j} + \beta_{ij,i(j+1)}x_{i(j+1)} +$$

$$\beta_{(i-1)j,ij}x_{(i-1)j} + \beta_{i(j-1),ij}x_{i(j-1)} \tag{4.20}$$

As we said earlier, temperature, T, has a very important role in Gibbs distribution. It controls the *convergence* of the distribution. In equation (4.18), as $T \to 0$, the value of $p(x_i = 1|N_i)$ tends to a limit depending on the W_i. If $W_i > 0$, then $p(x_i = 1|N_i) \to 0$ as $T \to 0$. Conversely, if $W_i < 0$, then $p(x_i = 1|N_i) \to 1$ as $T \to 0$. If $W_i = 0$, then $p(x_i = 1|N_i) = 0.5$ regardless of the value of T. Therefore, the W_i are indicators of whether the x_i at the position i should be 1 or -1. This indication becomes stronger as the temperature is cooled towards zero. DEUM algorithms exploit the temperature coefficient to balance the exploration and exploitation of the search space. Once marginals are defined from joint distribution, the next step in DEUM is to sample new solution. By its definition, a Markov network structure may contain cycles (apart from those that assume univariate structure). It is therefore not possible to use a Probabilistic logic sampling (PLS) approach as used by Bayesian network based EDAs. DEUM use Markov Chain Monte Carlo (MCMC) [17] methods for sampling. MCMC are a iterative sampling approach that does not require the ancestral ordering of variables. An instance of it is described later in section 5. Next we review three of the key DEUM algorithms proposed to date.

[5] For the purpose of arithmetical symmetry, in MFM a binary variable x_i is set to $\{-1, 1\}$ as opposed ot $\{0, 1\}$.

4.4 DEUM$_d$ - A Univariate DEUM Algorithm with Direct Sampling of Gibbs Distribution

In this section, we describe DEUM$_d$ [35, 36], a DEUM algorithm that assumes each variables in the solution to be independent. As such it fall in the class of univariate EDA. DEUM$_d$ is an update to the earlier proposed DEUM$_{pv}$ algorithm [34]. The workflow of DEUM$_d$ is shown in figure 4.3. It begins by initialising a population of solution P. Then the selection process takes place where N best solution is selected from P. MRF parameters, α, are then calculated by fitting the univariate MFM, (4.13), on the selected set of solution. As described in previous section, this is achieved by solving the system of linear equations (4.15). The $p(x_i)$ is then calculated from using α (as shown in figure 4.3) and sampled to generate the child population. The child then replaces the parent, P, and this process continues until termination criteria are satisfied.

DEUM$_d$ Workflow

1. Generate a population, P, consisting of M solutions
2. Select a set D from P consisting of N best solutions, where $N \leq M$.
3. For each solution, x, in D, build a linear equation of the form

$$-ln(f(x)) = \alpha_0 + \alpha_1 x_1 + \alpha_2 x_2 + ... + \alpha_n x_n$$

   Where, $f(x)$, the fitness of the solution x, should be ≥ 1; $\alpha = \{\alpha_0, \alpha_1, \alpha_2, ..., \alpha_n\}$ are equation parameters. Notice that α_0, the intercept of the linear equation, is added to the energy function. This is standard practice and helps to better fit the system of equations.
4. Solve the build system of N equations to estimate α
5. Use α to estimate univariate marginal probabilities as

$$p(x_i = 1) = \frac{1}{1 + e^{\beta \alpha_i}}, \quad p(x_i = -1) = \frac{1}{1 + e^{-\beta \alpha_i}}$$

   Here, β (inverse temperature coefficient) is set to $\beta = g \cdot \tau$; g is current iteration of the algorithm and τ is the parameter known as the cooling rate
6. Generate M new solution by sampling marginal probabilities to replace P and go to step 2 until termination criteria are meet

---

Fig. 4.3 The workflow of DEUM$_d$

As described earlier, β, the inverse temperature coefficient (in figure 4.3), has a direct effect on the convergence speed of DEUM$_d$. As the number of iterations, g, grows, the marginal probability, $p(x_i)$, gradually cools to either 0 or 1. However, depending upon the type of problem, different cooling rates may be required. In particular, there is a trade-off between convergence speed of the algorithm and the exploration of the search space. Therefore, the cooling rate parameter, τ, has been introduced. τ gives explicit control over the convergence speed of DEUM$_d$.

Decreasing τ slows the cooling, resulting in better exploration of the search space. However, it also slows the convergence of the algorithm. Increasing τ, on the other hand, makes the algorithm converge faster. However, the exploration of the search space will be reduced.

4.5 Is-DEUM$_g$ - A Multivariate DEUM with Ising Structure and Gibbs Sampler

In this section we describe Is-DEUM$_g$, a version of DEUM that use a fixed multivariate model (figure 4.2). Since, this model can be seen as a variant of Ising model [17], the prefix Is- has been used. As stated earlier, a Markov network structure (such as figure (4.2)) encode mutual dependencies between variables, and because there is no ancestral ordering of variables, sampling it becomes a difficult task. In order to extend DEUM to multivariate case, it is necessary to resolve this issue. In Is-DEUM$_g$, an iterative sampling technique, known as Markov Chain Monte Carlo samplers (MCMC) [17] is used. Specifically, we are interested in Gibbs sampler (GS) [9]: a well known instance of the MCMC sampler.

Before going further it is important to understand the concept behind Gibbs samplers, as this forms the basis for sampling techniques in other multivariate DEUM algorithms (as well as in many other Markov network based EDAs). Figure 4.4 shows the workflow of a Gibbs sampler.

Gibbs Sampler (GS)

1. Generate a solution $x^o = \{x_1^o, x_2^o, .., x_n^o\}$ and also set initial value for T.
2. Select a variable x_i^o and set $x_i^o = 1$ with probability $p(x_i^o = 1|N_i^o)$ (4.18)
3. Decrease T and go to 2 until a termination criteria is satisfied
4. Terminate with answer x^o.

Fig. 4.4 The pseudo-code for a Gibbs Sampler

The general idea here is to repeatedly sample variables in x^o (figure 4.4) until a termination criteria is satisfied, such that a (locally) optimal x^o is produced. Different termination criteria could be used for this purpose, for example, to terminate after a fixed number of iteration is performed, or to terminate if no further improvement in energy $U(x^o)$ could be found. The temperature coefficient, T, in Gibbs sampler can be used to control the convergence of $p(x_i^o|N_i^o)$. In each iteration, T is decreased. This gradually converges $p(x_i^o|N_i^o)$ to its limit. This iterative process would produce a x^o that, depending upon the allowed iteration, would be closer to the optima encoded by current set of MRF parameters θ.

Figure 4.5 shows the workflow of Is-DEUM$_g$, which incorporates Gibbs sampler as its sampling method.

Is-DEUM$_g$ workflow

1. Generate a population, P, of size M
2. Select the set D consisting of N fittest solutions from P, where $N \leq M$.
3. Calculate the MRF parameters θ by fitting (4.14) to D.
4. Run Gibbs sampler M times to generate new population.
5. If termination criteria is not satisfied, replace parent with new population and go to step 2

Fig. 4.5 Is-DEUM$_g$ workflow

As stated earlier, the convergence of GS depends on two factors: a) how fast the temperature T is decreased, and b) how many iteration the GS is allowed to do. This, therefore, also effects the performance of the Is-DEUM$_g$. Decreasing T quickly may result in premature convergence of x^o. Conversely, decreasing T slowly may result in high computation cost. Similarly, allowing GS to iterate for large number of runs would converge x^o to some local optima pointed by the current set of MRF parameters θ. This, therefore, would result in increasing number of similar solutions being present in the new population that are converged to some local optima. If the current optima pointed by θ is not the global optima for the problem, the result would be a quick loss of diversity in the population, even straight after the initial generation. Therefore, setting the correct rate of change for temperature and setting the allowed number of iteration is crucial in Is-DEUM$_g$. In [31], a DEUM with another MCMC sampling technique, known as Metropolis sampling [17], is also described.

4.6 Full DEUM - A Fully Multivariate General DEUM Algorithm

The two DEUM algorithms described in previous sections have one key similarity. They do not learn the structure of the problem and assume that it is known in advance. Here, we describe a fully multivariate DEUM algorithm [32] that can automatically learn the undirected structure of the problem, automatically find the cliques from the structure and automatically estimate a joint probability model of the Markov network. The workflow of full DEUM is is shown in Figure 4.6.

As can be seen, the estimation of undirected structure and building of a MFM (Step (3) and (4) in Figure 4.6) is the additional part in full DEUM algorithm that has not been implemented in other DEUM instances. Let us describe these key features of the full DEUM algorithm in more detail.

Distribution Estimation using Markov networks (DEUM)

1. Generate parent population P
2. Select a set of solutions D from P
3. Estimate undirected structure G of the Markov network from D
4. Find all the cliques in structure G and build a fitness model (MFM)
5. Estimate parameters of the Markov network by fitting build MFM to D
6. Sample Markov network to generate new solutions
7. Go to step 2 until termination criteria are meet

Fig. 4.6 The pseudo-code of the fully multivariate Distribution Estimation Using Markov network (DEUM) algorithm

4.6.1 Estimation of Undirected Structure

A number of different approaches can be used to estimate an undirected structure from data. In full DEUM a entropy based approach, initially described in [40] is used. More precisely, mutual information of each pairs of variable in the solution is estimated to create a matrix of mutual information. Mutual information between two random variables, A and B, is given by

$$I(A,B) = \sum_{a,b} p(a,b) log \left(\frac{p(a,b)}{p(a) \cdot p(b)} \right) \tag{4.21}$$

Here sum is over all possible combinations of A and B, and $p(a,b)$ is the joint probability of $A = a$ and $B = b$ computed from D. The pairs with the mutual information higher than a certain threshold TR are then made neighbours. The threshold, TR, is computed as $TR = avg(MI) * sig$, where $avg(MI)$ is the average of the elements of the mutual information matrix and sig is the significance parameter, useually set to 1.5. Also, in order to avoid an overly complex network, the number of neighbours that a variable can have is limited to a certain number. Note that other general statistical tests, such as chi square (also use by [1]), could also be used as the measurer for dependency as opposed to mutual information.

4.6.2 Finding Cliques and Assigning Potentials

Once the undirected structure is found, next step is to find all the cliques in the structure. For this, Bron-Kerbosch algorithm [3] has been used. We do not provide the details of this algorithm. Interested readers are suggested to see [3]. Once the set of maximal cliques are found, the potential functions are assigned to them. There are three possibilities while defining clique potentials.

1. Define potentials to all the maximum cliques
2. Define potentials to all the maximum cliques and all the subcliques[6].
3. Define potentials to all the maximum cliques and only some of the sub-cliques within it

The first option is the simplest and most efficient one, which requires only one potential function to be assigned per clique and, therefore require only one MRF parameter to be defined per clique. The second option is the safest one since it considers all possible sub interaction with the cliques. At the same time, it is the most expensive way to define MFM, since the number of MRF parameter grows significantly depending upon the order of the maximal cliques. The third option is the compromise between first and second, however, it may not be obvious to choose the sub-cliques required to build the model. Further research should be done in this area. In full DEUM presented in [32], the first option is used.

4.6.3 Sampling New Solution

Once the MFM is built, it is then fitted to the set of selected solution D (as described earlier in section (4.3.4)) and the MRF parameters are estimated. These MRF parameters fully specify the joint distribution (4.3), which is then sampled to generate new solutions. Similar to Is-DEUM$_g$, full DEUM use Gibbs sampler (Figure 4.4) for his purpose. In [32], another version of Gibbs sampler that does a raster scaning of bits is also described. Notice that each run of Gibbs sampler will sample a single solution. Running it multiple times will give a population of solution, which then replaces the parent population and the next iteration of DEUM follows.

4.7 Summary

DEUM is one of the early EDAs to use Markov Networks as its model of probability distribution. It uses undirected graph to represent variable interaction in the solution, and builds a model of fitness function from it. The model is then fitted to the set of solutions to estimate the network parameters; these are then sampled to generate new solutions. Over the years, many different DEUM algorithms have been proposed. In this chapter, we presented key components of DEUM algorithm and described the workflow of some of the well known instances of it.

DEUM algorithms have been applied to many benchmark and real world problems. They include Schaffer f6[34], Plateau [36], Checkerboard problem [36], Equal products function [36], Colville [36], Trap [31], SixPeaks function [36], Mushroom farming [16], Revenue management/Dynamic pricing [38, 39], Cancer Chemo Therapy [15], Ising Spinn glasses [31], MaxSat [5]. Their results were shown to be

[6] Subcliques are defined as the smaller cliques within the maximal cliques. For example, in Figure 4.1, variables $\{X_1, X_2\}$ defines a sub-clique within clique $\{X_1, X_2, X_3\}$. Furthermore, variable $\{X_1\}$ also defines a singleton sub-clique within $\{X_1, X_2, X_3\}$.

comparable and at times better than other EDAs of their class. Chapter 12 in this book reviews some of the interesting real world applications of DEUM algorithm.

One of the key distinctions between DEUM and other Markov network based EDA is its fitness modelling approach to parameterising Markov networks. It builds a model of fitness function that fully specifies the joint distribution. This is then sampled using Markov Chain Monte-Carlo samplers. The fitness modelling can be a powerful tool to enhance the performance of an optimisation algorithm. The complex fitness function can be modelled to a simplified linear (or non linear) models. Fitness modelling techniques have also been applied in surrogate fitness models [14, 21, 30], and guided genetic operators [22, 43–45]. Apart from fitness modelling, another key component of DEUM algorithm is the use of Gibbs sampler technique to sample the network. The temperature coefficient in Gibbs sampler gives DEUM an explicit control over the exploration and exploitation of the search space. Finally, the automated structure learning and model building techniques incorporated in full DEUM gives these algorithms an increased versatility that can be useful to solve difficult real world problems.

References

1. Alden, M.A.: MARLEDA: Effective Distribution Estimation Through Markov Random Fields. PhD thesis, Faculty of the Graduate Schoool, University of Texas at Austin, USA (December 2007)
2. Besag, J.: Spatial interactions and the statistical analysis of lattice systems (with discussions). Journal of the Royal Statistical Society 36, 192–236 (1974)
3. Born, C., Kerbosch, J.: Algorithms 457 - finding all cliques of an undirected graph. Communications of the ACM 16(6), 575–577 (1973)
4. Brown, D.F., Garmendia-Doval, A.B., McCall, J.A.W.: Markov Random Field Modelling of Royal Road Genetic Algorithms. In: Collet, P., Fonlupt, C., Hao, J.-K., Lutton, E., Schoenauer, M. (eds.) EA 2001. LNCS, vol. 2310, pp. 65–78. Springer, Heidelberg (2002)
5. Brownlee, A., McCall, J., Brown, D.: Solving the MAXSAT problem using a Multivariate EDA based on Markov Networks. In: A Late Breaking Paper in GECCO 2007: Proceedings of the 2007 Conference on Genetic and Evolutionary Computation, Global Link Publishing (2007)
6. Brownlee, A., McCall, J., Zhang, Q., Brown, D.: Approaches to selection and their effect on fitness modelling in an estimation of distribution algorithm. In: Proceedings of the 2008 Congress on Evolutionary Computation CEC 2008, pp. 2621–2628. IEEE Press, Hong Kong (2008)
7. Brownlee, A.E.I.: Multivariate Markov networks for fitness modelling in an estimation of distribution algorithm. PhD thesis, The Robert Gordon University. School of Computing, Aberdeen, UK (2009)
8. Etxeberria, R., Larrañaga, P.: Global optimization using Bayesian networks. In: Ochoa, A., Soto, M.R., Santana, R. (eds.) Proceedings of the Second Symposium on Artificial Intelligence (CIMAF 1999), Havana, Cuba, pp. 151–173 (1999)

9. Geman, S., Geman, D.: Stochastic relaxation, Gibbs distributions and the Bayesian restoration of images. In: Fischler, M.A., Firschein, O. (eds.) Readings in Computer Vision: Issues, Problems, Principles, and Paradigms, pp. 564–584. Kaufmann, Los Altos (1987)
10. Hammersley, J.M., Clifford, P.: Markov fields on finite graphs and lattices (1971) (unpublished)
11. Kindermann, R., Snell, J.L.: Markov Random Fields and Their Applications. AMS (1980)
12. Larrañaga, P., Etxeberria, R., Lozano, J.A., Peña, J.M.: Combinatorial optimization by learning and simulation of Bayesian networks. In: Proceedings of the Sixteenth Conference on Uncertainty in Artificial Intelligence, Stanford, pp. 343–352 (2000)
13. Li, S.Z.: Markov Random Field modeling in computer vision. Springer (1995)
14. Lim, D., Jin, Y., Ong, Y.-S., Sendhoff, B.: Generalizing surrogate-assisted evolutionary computation. IEEE Transactions on Evolutionary Computation (2008)
15. McCall, J., Petrovski, A., Shakya, S.: Evolutionary algorithms for cancer chemotherapy optimisation. In: Fogel, G., Corne, D., Pan, Y. (eds.) Computational Intelligence in BioInformatics. Wiley, Chichester (2007)
16. McCall, J., Wu, Y., Godley, P., Brownlee, A., Cairns, D., Cowie, J.: Optimisation and fitness modelling of bio-control in mushroom farming using a markov network eda. In: Proceedings of Genetic and Evolutionary Computation Conference (GECCO 2008). ACM (2008)
17. Metropolis, N.: Equations of state calculations by fast computational machine. Journal of Chemical Physics 21, 1087–1091 (1953)
18. Mühlenbein, H., Mahnig, T.: FDA - A scalable evolutionary algorithm for the optimization of additively decomposed functions. Evolutionary Computation 7(4), 353–376 (1999)
19. Murphy, K.: Dynamic Bayesian Networks: Representation, Inference and Learning. PhD thesis, University of California, Berkeley (2002)
20. Murray, I., Ghahramani, Z.: Bayesian Learning in Undirected Graphical Models: Approximate MCMC algorithms. In: Twentieth Conference on Uncertainty in Artificial Intelligence (UAI 2004), Banff, Canada, July 8-11 (2004)
21. Ong, Y.S., Nair, P.B., Keane, A.J., Wong, K.W.: Surrogate-assisted evolutionary optimization frameworks for high-fidelity engineering design problems. In: Knowledge Incorporation in Evolutionary Computation, pp. 307–332. Springer (2004)
22. Peña, J.M., Robles, V., Larrañaga, P., Herves, V., Rosales, F., Pérez, M.S.: GA-EDA: Hybrid Evolutionary Algorithm Using Genetic and Estimation of Distribution Algorithms. In: Orchard, B., Yang, C., Ali, M. (eds.) IEA/AIE 2004. LNCS (LNAI), vol. 3029, pp. 361–371. Springer, Heidelberg (2004)
23. Pearl, J.: Probabilistic Reasoning in Intelligent Systems. Morgan Kaufman Publishers, Palo Alto (1988)
24. Pelikan, M.: Bayesian optimization algorithm: From single level to hierarchy. PhD thesis, University of Illinois at Urbana-Champaign, Urbana, IL, Also IlliGAL Report No. 2002023 (2002)
25. Pelikan, M., Goldberg, D.E.: Hierarchical problem solving by the Bayesian optimization algorithm. IlliGAL Report No. 2000002, Illinois Genetic Algorithms Laboratory, University of Illinois at Urbana-Champaign, Urbana, IL (2000)
26. Press, W.H., Teukolsky, S.A., Vetterling, W.T., Flannery, B.P.: Numerical Recipes in C: The Art of Scientific Computing, 2nd edn. Cambridge University Press, Cambridge (1993)

27. Santana, R.: A Markov Network Based Factorized Distribution Algorithm for Optimization. In: Lavrač, N., Gamberger, D., Todorovski, L., Blockeel, H. (eds.) ECML 2003. LNCS (LNAI), vol. 2837, pp. 337–348. Springer, Heidelberg (2003)
28. Santana, R.: Estimation of Distribution Algorithms with Kikuchi Approximations. Evolutionary Computation 13, 67–98 (2005)
29. Santana, R., Larrañaga, P., Lozano, J.A.: Mixtures of Kikuchi Approximations. In: Fürnkranz, J., Scheffer, T., Spiliopoulou, M. (eds.) ECML 2006. LNCS (LNAI), vol. 4212, pp. 365–376. Springer, Heidelberg (2006)
30. Sastry, K., Lima, C., Goldberg, D.E.: Evaluation relaxation using substructural information and linear estimation. In: Proceedings of the Genetic and Evolutionary Computation Conference (GECCO 2006), pp. 419–426. ACM Press, New York (2006)
31. Shakya, S.: *DEUM: A Framework for an Estimation of Distribution Algorithm based on Markov Random Fields*. PhD thesis, The Robert Gordon University, Aberdeen, UK (April 2006)
32. Shakya, S., Brownlee, A., McCall, J., Fournier, F., Owusu, G.: DEUM – A Fully Multivariate EDA Based on Markov Networks. In: Chen, Y.-p. (ed.) Exploitation of Linkage Learning. Adaptation, Learning, and Optimization, vol. 3, pp. 71–93. Springer, Heidelberg (2010)
33. Shakya, S., McCall, J.: Optimisation by Estimation of Distribution with DEUM framework based on Markov Random Fields. International Journal of Automation and Computing 4, 262–272 (2007)
34. Shakya, S., McCall, J., Brown, D.: Updating the probability vector using MRF technique for a univariate EDA. In: Onaindia, E., Staab, S. (eds.) Proceedings of the Second Starting AI Researchers' Symposium. Frontiers in Artificial Intelligence and Applications, vol. 109, pp. 15–25. IOS press, Valencia (2004)
35. Shakya, S., McCall, J., Brown, D.: Estimating the distribution in an EDA. In: Ribeiro, B., Albrechet, R.F., Dobnikar, A., Pearson, D.W., Steele, N.C. (eds.) Proceedings of the International Conference on Adaptive and Natural Computing Algorithms (ICANNGA 2005), pp. 202–205. Springer, Wien (2005)
36. Shakya, S., McCall, J., Brown, D.: Using a Markov Network Model in a Univariate EDA: An Emperical Cost-Benefit Analysis. In: Proceedings of Genetic and Evolutionary Computation Conference (GECCO 2005), pp. 727–734. ACM, Washington, D.C (2005)
37. Shakya, S., McCall, J., Brown, D.: Solving the Ising spin glass problem using a bivariate EDA based on Markov Random Fields. In: Proceedings of IEEE Congress on Evolutionary Computation (IEEE CEC 2006), pp. 3250–3257. IEEE Press, Vancouver (2006)
38. Shakya, S., Oliveira, F., Owusu, G.: An application of GA and EDA to Dynamic Pricing. In: Proceedings of Genetic and Evolutionary Computation Conference (GECCO 2007), pp. 585–592. ACM, London (2007)
39. Shakya, S., Oliveira, F., Owusu, G.: Analysing the effect of demand uncertainty in dynamic pricing with eas. In: Bramer, M., Coenen, F., Petridis, M. (eds.) Research and Development in Intelligent Systems XXV: Proceedings of AI 2008, The Twenty Eighth SGAI International Conference on Innovative Techniques and Applications of Artificial Intelligence, Cambridge, UK, pp. 77–90. Springer, London (2008)
40. Shakya, S., Santana, R.: An EDA based on local Markov property and Gibbs sampling. In: Proceedings of Genetic and Evolutionary Computation Conference (GECCO 2008), Atlanta, Georgia, USA. ACM (2008)
41. Shakya, S., Santana, R.: A markovianity based optimisation algorithm. Technical Report Technical Report EHU-KZAA-IK-3/08, Department of Computer Science and Artificial Intelligence, University of the Basque Country (September 2008)

42. Shakya, S., Santana, R.: A markovianity based optimisation algorithm. Genetic Programming and Evolvable Machines (2011) (in press)
43. Sun, J., Zhang, Q., Li, J., Yao, X.: A hybrid estimation of distribution algorithm for cdma cellular system design. International Journal of Computational Intelligence and Applications 7(2), 187–200 (2008)
44. Zhang, Q., Sun, J.: Iterated local search with guided mutation. In: Proceedings of the IEEE World Congress on Computational Intelligence (CEC 2006), pp. 924–929. IEEE Press (2006)
45. Zhang, Q., Sun, J., Tsang, E.: An evolutionary algorithm with guided mutation for the maximum clique problem. IEEE Transactions on Evolutionary Computation 9(2), 192–200 (2005)

Chapter 5
MN-EDA and the Use of Clique-Based Factorisations in EDAs

Roberto Santana

Abstract. This chapter discusses the important role played by factorisations in the study of EDAs and presents the Markov network estimation of distribution algorithm (MN-EDA) as a classical example of the EDAs based on the use of undirected graphs. The chapter also reviews recent work on the use of clique-based decompositions and other approximations methods inspired in the field of statistical physics with direct application to EDAs.

5.1 Introduction

The question of problem decomposition is ubiquitous in machine learning research. Problem solving can be considerably simplified in a principled way to decompose the original problem in simpler subproblems. The idea of problem decomposition was also at the very core of the original conception of GAs. Parallel combination of partial solutions (or building blocks) is one of the explanations for the success of GAs. Therefore, an impressive amount of work has been devoted to the conception of methods for automatic identification and efficient combination of partial solutions. In EDAs, the question of problem decomposition can be translated into the problem of finding effective and efficient factorisations.

Constructing feasible probability distribution approximations is one of the most relevant issues in EDAs. The effectiveness of a probability approximation can be linked in these algorithms to its capacity for capturing the wide variety of the complex relationships that may arise between the problem variables during the optimisation process. The feasibility of the approximation also depends on the complexity of the methods used to learn the models from the selected set and the procedures used to sample the new solutions. In general, probability models are expected to be learnt

Roberto Santana
Intelligent Systems Group, Faculty of Informatics,
University of the Basque Country (UPV/EHU), San-Sebastian, Spain
e-mail: roberto.santana@ehu.es

S. Shakya and R. Santana (Eds.): Markov Networks in Evolutionary Computation, ALO 14, pp. 73–87.
springerlink.com

without an excessive computational burden and the implementation of the sampling process should be simple enough to allow the generation of several individuals in each generation, as is the general rule in EDAs.

The traditional approach to problem solving using EDAs explicitly or implicitly determines a choice of a class of probabilistic graphical model (PGM), the one that is used by the selected EDA. The performance of the algorithm is evaluated in terms of the optimisation results. Therefore, only rarely the capacity of the PGM to accurately modelling the selected individuals is measured. When the EDA fails to solve the problem, it is usually assumed, as a possible explanation for the results, that the chosen PGM is not able to accurately model the characteristics of the selected individuals. EDAs that use different PGMs are then tried until a suitable candidate algorithm is found. However, deep knowledge about EDA behaviour and expertise in model selection for optimisation requires a better understanding of the relationship between the choice of the probabilistic models and the performance of the EDA. The concept of factorisation is key for this purpose.

A factorisation of a distribution is a representation of the distribution as a product of marginal and conditional distributions, the factors, which are defined on subsets of the complete set of variables. Factorisations are important because they allow us to obtain a condensed representation of otherwise very difficult to store probability distributions. Different combinations of factors may be seen as alternative ways to solve the puzzle of an efficient optimisation. How to identify the pieces of the puzzle and organise an efficient strategy to assemble it is the problem underlying the steps of learning and sampling factorisations.

A first attempt to classify factorisations according to some characteristics of the graphical models was presented in [11]. In this paper, Mühlenbein et al. proposed to separate factorisations into two classes (valid vs invalid), in accordance to the fulfilment of the running intersection property. In [15], factorisations were further analysed and the classification between ordered and messy factorisations was introduced as a way to ease the identification and construction of feasible probability distribution approximations. As an example of algorithms that produce messy factorisations, the Markov network estimation of distribution algorithm (MN-EDA) was introduced. MN-EDA exhibits a number of characteristics different to previous EDAs. It learns a probability approximation starting from a Markov network. To construct a feasible factorisation, it applies a method inspired in the use of region-based approximations of the free energy, as used in statistical physics [22]. Finally, it uses a sampling procedure based on Gibbs sampling which allows the incorporation of a priori information about the problem. The numerical results presented in [15] showed that MN-EDA is able to solve very difficult deceptive and hierarchical problems by means of representing complex interactions among the variables of the problem.

MN-EDAs is a paradigmatic example of EDAs that construct invalid factorisations and therefore requires sampling methods different to those traditionally applied. Work on MN-EDA also illustrates the important role played by factorisations in the understanding and improvement of EDAs. In this chapter, we review the rationale behind the conception of MN-EDA and explain in detail its main components. Recent related developments in this research trend are also described.

The remainder of this chapter is presented as follows. The next section discusses the use of factorisations in EDAs. Section 5.3 introduces the concept of a clique based decomposition of a graph and presents the Kikuchi approximation of the probability distribution. Sections 5.4 and 5.5 respectively introduce a procedure for sampling points and an algorithm for learning Kikuchi approximations from the data. Section 5.6 describes MN-EDA. Section 5.7 reviews recent results and work related to the use of Kikuchi approximations and other clique-based approximations in EDAs. The conclusions of the chapter are presented in Section 5.8.

5.2 Factorisations and Factorised Distribution Algorithms

Let $X = (X_1,\ldots,X_n)$ denote a vector of discrete random variables. We will use $x = (x_1,\ldots,x_n)$ to denote an assignment to the variables. S will denote a set of indices in $\{1,\ldots,n\}$, and X_S (respectively x_S) a subset of the variables of X (respectively x) determined by the indices in S. We will work with positive distributions denoted by p. $p(x_S)$ will denote the marginal probability for $X_S = x_S$. We use $p(x_i \mid x_j)$ to denote the conditional probability distribution of $X_i = x_i$ for given $X_j = x_j$.

In simple terms, a factorisation of a distribution p is a product of marginals of p. We say that a factorisation is valid when the maximal cliques in the factorisation satisfy the running intersection property (RIP). Otherwise the factorisation is said to be invalid. As presented in Chapter 2, fulfilment of RIP implies that any two cliques containing a node α are either adjacent in the junction tree or connected by a chain made entirely of cliques that contain α.

There exist methods for finding valid factorisations based on the independence graph. It is very simple to construct an invalid factorisation, but we can not guarantee the accuracy of an approximation of the distribution based on an arbitrary, invalid factorisation. A criterion for the selection of invalid factorisations is needed as is a more refined classification of the overly broad and ill defined class of invalid factorisations.

In [14], a method for constructing invalid factorisations has been proposed. The method consists of constructing a labelled junction graph from the maximal cliques of the independence graph of the distribution. Even if the factorisation is invalid, it exhibits a convenient feature: the cliques in the factorisation can be ordered in such a way that at least one of the variables in every clique is not contained in the previous nodes in the ordering. This property is crucial for the implementation of the sampling procedure as explained in [14]. The factorisation based on the junction graph serves to propose a second classification of factorisations. We say that a factorisation is "ordered" [15] when an ordering of all the maximal cliques in the factorisation exists such that for every clique there exists at least one variable that is not contained in the previous cliques in the ordering. When it is impossible to find such an ordering, we say that the factorisation is "messy". The satisfaction of the running intersection property determines that every valid factorisation is an ordered factorisation.

Example 1 describes valid and invalid factorisations constructed from different independence graphs. The relationship between different classes of factorisations is shown in Figure 5.2.

Example 1. In this example, p_0 is an ordered, valid factorisation. The cliques where marginals are calculated from, can be joined to form a junction tree. p_1 is an ordered, invalid factorisation. In this case, cliques can not be arranged in a junction clique, but a labelled junction graph can be formed.

$$p_0(x_1,x_2,x_3,x_4,x_5,x_6) = \frac{p(x_1,x_2,x_3) \cdot p(x_1,x_2,x_5) \cdot p(x_1,x_3,x_6) \cdot p(x_2,x_3,x_4)}{p(x_1,x_2) \cdot p(x_1,x_3) \cdot p(x_2,x_3)}$$

$$p_1(x_1,x_2,x_3,x_4,x_5,x_6) = \frac{p(x_1,x_2,x_3) \cdot p(x_1,x_2,x_5) \cdot p(x_1,x_3,x_6) \cdot p(x_4,x_5,x_6)}{p(x_1,x_2) \cdot p(x_1,x_3) \cdot p(x_5) \cdot p(x_6)}$$

Figure 5.1 shows the graphical models from which these approximations have been constructed.

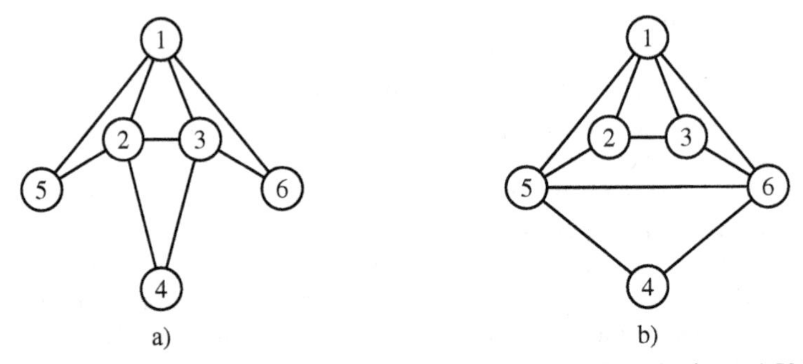

Fig. 5.1 Independence graphs illustrating different examples of factorisations. a) Valid factorisation used. b) Invalid factorisation.

In [15], the class of EDAs that use ordered factorisations of the probability distribution are called factorised distribution algorithms (FDAs). However, in the literature, the term FDA is frequently used to name a particular type of FDAs introduced in [11]. We use the definition of FDAs as introduced in [15]. The Markov Network FDA (MN-FDA) [14] extends the representation capabilities of previous FDAs. The main difference with respect to previous FDAs based on undirected models is that it uses a junction graph as its probabilistic model, being able to represent ordered invalid factorisations.

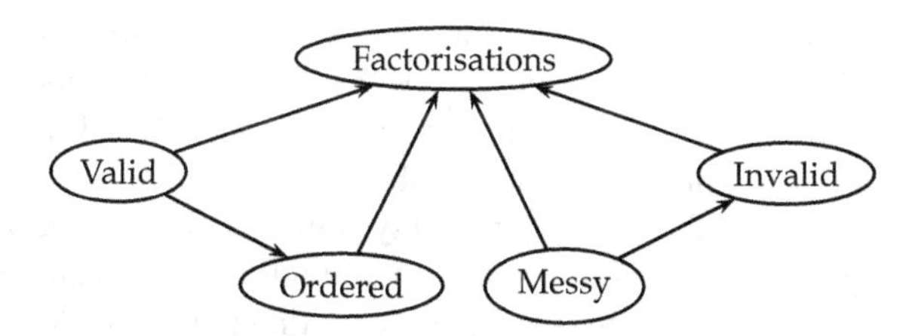

Fig. 5.2 Classifications for different types of factorisations

5.3 Region-Based Approximations of a Probability Distribution

This section deals with the question of constructing invalid "messy" factorisations. The approximation achieved by the method can be considered itself as an approximation of a Gibbs field.

Let us suppose that the structure of the Markov network is available, and that approximate marginal probabilities for every maximal clique of the graph are also known. Based on this information, we would study how to discover an approximate probability distribution that captures the information in the data.

To learn an Markov network, it is not enough to know the cliques potentials, it is also necessary to calculate the partition function Z. In general, finding Z is infeasible because it implies the summation on an exponential number of states. Therefore, if we intend to learn an MRF, a feasible solution to the problem of approximating Z has to be given. Fortunately, a similar problem has been treated in the field of statistical mechanics.

Statistical mechanics often treats models of interacting many-body systems. The macroscopic properties of the system are determined by the interactions among the microscopic elements. Usually, the interactions among the components of the system also determine the likelihood of the corresponding global configuration. Each state x of the system has an associated energy $H(x)$. A fundamental result of statistical mechanics is that, in thermal equilibrium, the probability of a state will be given by Boltzmann's Law:

$$p(x) = \frac{1}{Z(T)} e^{\frac{-H(x)}{T}}$$

where T is the temperature of the system, and $Z(T)$ the corresponding partition function.

Yedidia et al. [22] suggested using the Boltzmann law postulate to define the energy corresponding to a joint probability of a non-physical system. The temperature is arbitrarily set to 1 because it simply sets a scale for the units in which one measures the energy. Similar proposals have been successfully used in optimisation [7, 11].

Different techniques can be used to obtain an approximation of the energy. Kikuchi (1951) [6] developed the Cluster Variation Method (CVM). At the base of CVM is the idea of approximating the energy as a function of a set of different marginals. The set does not necessarily form an exact factorisation of the distribution, and the choice of the marginals influences the quality of the approximation.

We concentrate on the question of finding an appropriate set of marginals from which to obtain an approximation of the distribution.

We define a region R of the neighbourhood system $G = (V, E)$ to be a set $V' \subset V$. A graph region based decomposition is an asset of regions $\mathscr{R}$, and an associated set of *counting numbers* U which is formed by one counting number c_R for each $R \in \mathscr{R}$. c_R will always be an integer, but might be zero or negative for some R.

In [15], The Kikuchi approximation of a probability is defined on a graph region based decomposition, calculating the marginal probabilities for each region and combining them in a similar way to the way CVM is applied to free energy approximations. The methods used for creating the region based decomposition of the graph determine the final Kikuchi approximation. To be valid, a decomposition must satisfy a number of constraints related to the regions and the counting numbers. Inspired in the work by Yedidia et al. [23], we refer to this sub-problem as *finding a valid region based decomposition of a graph*. In the CVM, $\mathscr{R}$ is formed by an initial set of regions $\mathscr{R}_0$ such that all the nodes are in at least one region of $\mathscr{R}_0$, and any other region in $\mathscr{R}$ is the intersection of one or more of the regions in $\mathscr{R}$. The set of regions $\mathscr{R}$ is closed under intersection, and can be ordered as a poset.

We say that a set of regions $\mathscr{R}$, and counting numbers c_R give a valid region based graph decomposition when for every variable X_i:

$$\sum_{\substack{R \in \mathscr{R} \\ X_i \subseteq X_R}} c_R = 1 \tag{5.1}$$

This condition arises from the fact that acceptable approximations of the entropy component of the free energy are those in which the appearance of each variable in the set of regions sums 1 [12].

Yedidia et al. [23] introduced the concept of regions to define a region based free energy approximation on factor graphs. The key aspect in the determination of a valid region based decomposition is the selection of the initial set of regions. In [15], given an arbitrary undirected graph G, the set $\mathscr{R}_0$ is formed by taking one region for each maximal clique in G. As a result, all the regions $R \in \mathscr{R}$ will be cliques because they are the intersection of two or more cliques. We call this type of region based decomposition of undirected graphs a *clique based decomposition*. Alternative ways for selecting the initial regions are reported in [3, 5].

5.3.1 Kikuchi Approximation

The Kikuchi approximation of the probability, denoted as k is defined as [15]:

$$k(x) = \prod_{R \in \mathscr{R}} p(x_R)^{c_R}, \tag{5.2}$$

where $\mathscr{R}$ comes from a region based decomposition of the graph. From now on we will constrain our analysis to Kikuchi approximations of the probability constructed from clique based graph decompositions.

A probability distribution p can be used in two main different forms. As an evaluator, or as a generator. When p is used as an evaluator it associates a real probability value to a given point x. If we want to use p as a generator we will expect it to generate the point x with a probability $p(x)$. In this case, the outcome of p is not a probability value but a sample of vectors with the desired frequency. To use the Kikuchi approximation as an evaluator of a given point x, the marginals defined on each region are multiplied, or divided, according to the sign of c_R.

A probability function $\tilde{p}$ based on the Kikuchi approximation can be found by normalising k.

$$\tilde{p}(x) = \frac{k(x)}{\sum_{x'} k(x')}$$

However, the determination of $\tilde{p}$ is not a realistic alternative, because it does not solve the problem of calculating a partition function $Z_k = \sum_{x'} k(x')$.

The Kikuchi approximation has a number of properties that can be used in the context of EDAs [15]. In certain cases the Kikuchi approximation is a valid factorisation. It also satisfies the local Markov property.

5.3.2 Algorithm for Constructing Valid Region Based Decompositions

The construction of a clique based decomposition involves the determination of the overlapping regions and their associated counting numbers that will be used in the factorisation. Figure 5.3 presents the steps of the algorithm for finding a valid region based decomposition. The algorithm receives as input the first set of regions $\mathscr{R}_0$. The final decomposition has only one set of regions $\mathscr{R}_1$ whose c_R values, $R \in \mathscr{R}_1$, are integers different from zero.

Let $\mathscr{A}$ be an auxiliary set of regions, and $\Phi(\mathscr{A})$ be a function that finds all the intersection regions in $\mathscr{A}$. First, a candidate set of regions $\mathscr{Q}$ is constructed by progressively adding intersections of regions. When all possible intersections have been added, $\mathscr{Q}$ is ordered. The order is determined by the inclusion criterion. If $R_i \supseteq R_j$ then R_i precedes R_j in the ordering. Afterwards, each region R in $\mathscr{Q}$ is inspected, and its c_R value is calculated as

$$c_R = 1 - \sum_{\substack{S \in \mathscr{R}_1 \\ S \supset R}} c_S \tag{5.3}$$

where c_S is the counting number of any region S in $\mathscr{R}_1$ such that S is a superset of R. If c_R is different from zero, the region is added to $\mathscr{R}_1$.

All the maximal cliques in $\mathscr{R}_0$ will be in $\mathscr{R}_1$ with $c_R = 1$. The algorithm finds a partially ordered set of all the regions. It is guaranteed [10] that when such a poset

Valid region based decomposition

1. $\mathscr{A} = \mathscr{R}_0$
2. $\mathscr{Q} = \mathscr{A}$
3. **do**
4. $\quad \mathscr{A} = \Phi(\mathscr{A})$
5. $\quad \mathscr{Q} = \mathscr{Q} \cup \mathscr{A}$
6. **until** $\mathscr{A} = \emptyset$
7. Find a partial ordering of $\mathscr{Q}$
8. $\mathscr{R}_1 = \emptyset$
9. $U = \emptyset$
10. **for** $R \in \mathscr{Q}$
11. $\quad c_R = 1 - \sum_{\substack{S \in \mathscr{R}_1 \\ S \supset R}} c_S$
12. $\quad$ if $c_R \neq 0$ then
13. $\quad\quad \mathscr{R}_1 = \mathscr{R}_1 \cup R$
14. $\quad\quad U = U \cup c_R$

Fig. 5.3 Algorithm for constructing a valid region based decomposition

is closed under the intersection of regions, and the counting numbers are calculated using (5.3), the region based decomposition is valid.

5.4 Gibbs Sampling of the Kikuchi Approximation

To use the Kikuchi approximation as a generator, we will take advantage of the local Markov property it satisfies. As k is not a probability distribution, we assume that points will be generated from its normalised expression $\tilde{p}(x)$. To sample points from $\tilde{p}(x)$, we will use the Gibbs sampler or Gibbs Sampling algorithm [2]. It is assumed that whilst $\tilde{p}(x)$ is too complex to draw samples from it directly, its conditional distributions $\tilde{p}(x_i \mid x \setminus x_i) = k(x_i \mid x \setminus x_i)$ are tractable to work with.

To draw samples from $\tilde{p}(x)$, start with an initial state $x^0 = (x_1^0, \ldots, x_n^0)$. At each time t of the GS, a location i to be updated is chosen from x^t, and the new vector x^{t+1} is selected using the transition rules in (5.4).

$$x^{t+1} = \begin{cases} x_j^{t+1} = x_j^t & \text{for } j \neq i \\ x_j^{t+1} \approx k(x_i^t \mid bd(x_i^t)) & \text{for } j = i \end{cases} \tag{5.4}$$

VS, Cy, and In are the parameters of the GS algorithm. VS is the type of visitation scheme, and defines the way in which the variables are selected for update. Random ($VS = 0$), or fixed ($VS = 1$) visitation schemes can be used. Cy is the number of cycles of the GS algorithm. One cycle comprises the update of n variables. In is a parameter that determines the way the initial vector of the GS is constructed. The vector where the GS sampling starts from can be randomly selected ($In = 0$), or sampled from an approximate factorisation found using a chordal subgraph of the independence graph ($In = 1$). In [15], the influence of these parameters was investigated. It was shown that the behaviour of MN-EDA critically improves depending on an appropriate selection of these parameters.

5.5 Learning the Kikuchi Approximation from the Data

To use the Kikuchi approximation in the context of an EDA we need a method that learns the structure and parameters of the model. The methods used to learn the structure is essentially an algorithm for learning an undirected graphical model from data. It is based on the application of independence tests as proposed by Spirtes et al. [20] but adds a second step in which the maximal cliques in the graph are identified. Figure 5.4 presents the steps of the algorithm which is an an extension of the algorithm introduced in [14] for learning ordered factorisations from the data. Steps from 1 to 3 are the same, in steps 4 and 5 the Kikuchi approximation is constructed.

Learning the Kikuchi approximation

1. Learn an independence graph G from the data (the selected set of solutions).
2. If necessary, refine the graph.
3. Find the set $\mathscr{C}$ of all the maximal cliques of G.
4. Construct a clique based decomposition of the graph.
5. Find the marginal probabilities for the regions of the decomposition.

Fig. 5.4 Algorithm for learning the Kikuchi approximation from the data

The independence graph is constructed by means of independence tests. To determine if an edge belongs to the graph, it is enough to make an independence test on each pair of variables given the rest. Nevertheless, from an algorithmic point of view it is important to reduce the cost of the independence tests. We have adopted the methodology previously followed by [20]. The idea is to start from a complete undirected graph, and then try to remove edges by testing for conditional independence between the linked nodes, but using conditioning sets as small as possible.

As the order of tests increases their reliability decreases, and therefore the use of higher order tests is not practical, unless a big population size is employed. In [15], the Chi-square independence test is used. If two variables X_i and X_j are dependent with a specified level of significance α, they are joined by an edge. α is a parameter of the algorithm. The algorithm weights each edge $i \sim j$ in the independence graph with a value $w(i,j)$ stressing the pairwise interaction between the variables. We use the value of the Chi-square test to set $w(i,j)$.

When the independence graph is very dense, we can expect that the dimension of the cliques will increase beyond a feasible limit. It is important to impose a limit r to the size of the maximum clique. One solution to this problem is to make the graph sparser in one step previous to the calculation of the cliques. This has been done by allowing a maximum number $r-1$ of incident edges to each vertex. If one vertex has more than $r-1$ incident edges, those with the lowest weights are removed. In this way, the size of the maximum clique will always be smaller or equal than r.

To find all the cliques of the graphs the Bron and Kerbosch algorithm [1] is used. This algorithm uses a branch and bound technique to cut off branches that can lead to cliques. Once all the cliques have been found, they are stored in a list L, and their weights are calculated from the information about dependencies. The weight of any subgraph G' of G is calculated as $W(G') = \sum_{i \sim j \in G'} w(i,j)$. These weights associated with each maximal clique are used to construct a junction graph that serves to generate non-random initial points for the GS. The algorithm to construct the junction graph is explained in [14].

5.6 The Markov Network EDA

MN-EDA is related to MN-FDA because both algorithms share part of the procedure used for learning the probabilistic model. Nonetheless, while MN-FDA can only handle ordered factorisations, MN-EDA is able to use the messy ones. The goal of using the Kikuchi approximation in the context of EDAs is to have a probabilistic model able to capture as many dependencies as possible, but without adding new edges to the independence graph, as other procedures (e.g. triangulation algorithms) do. The MN-EDA's pseudo-code is shown in Figure 5.5.

Markov Network EDA

1. Set $t \Leftarrow 0$. Generate N solutions randomly.
2. **do**
3. Select a set S of $l \leq N$ solutions according to a selection method.
4. Learn a Kikuchi approximation of S using algorithm 5.4
5. Generate N new solutions sampling from the Kikuchi approximation.
6. $t \Leftarrow t+1$
7. **until** Termination criteria are met

Fig. 5.5 Markov Network estimation of distribution algorithm

In the implementation presented in [15], the level of significance α was set to 0.75. This choice was motivated by the need to capture as many dependencies as possible. A low significance level will allow more dependencies to be detected in the last generations, when many variables tend to be independent. Even if some of the dependencies found might be false, this is better than missing some of the real dependencies. If many dependencies are learnt, the refinement algorithm will contribute to eliminate those with lowest weights, balancing the effect of the low significance level. The maximum number of allowed neighbours for the refinement algorithm was set to 9.

The name MN-EDA^f is given to the MN-EDA that uses a fixed model of interactions. This algorithm can be given the clique based decomposition, or can calculate it from a given independence graph. Once the clique based decomposition has been constructed, it will be used in all the generations of the algorithm. In every generation, MN-EDA^f only makes a parametric approximation of the marginal probabilities of the regions.

5.7 Related Work and Extensions to MN-EDA

A number of extensions have been proposed to the use of clique-based decompositions in EDAs. In this section we review some of this recent work.

5.7.1 Mixture of Kikuchi Approximations

In [16], it was analysed how to enhance capacity of mixtures for probabilistic modelling considering Kikuchi approximations as the components of the mixture probability distributions. The idea was to create probabilistic models able to represent a higher number of dependencies of the data. The mixture of Kikuchi approximations EDA (MKA-EDA) was introduced and evaluated on the satisfiability problem. The rationale behind MKA-EDA was to combine the capacity of mixtures to exploit asymmetric independence assertions with the power of Kikuchi approximations to represent complex interactions.

A mixture of Kikuchi approximations [16] is defined to be an approximation of the form:

$$k^m(\mathbf{x}) = \sum_{j=1}^{m} \lambda_j k_j(\mathbf{x}) \tag{5.5}$$

with $\lambda_j > 0$, $j = 1,\ldots,m$, $\sum_{j=1}^{m} \lambda_j = 1$.

The components of $k^m(\mathbf{x})$ are Kikuchi approximations. Since Kikuchi approximations are not probability distributions in general, the mixture of Kikuchi approximations is not either. However, whenever the clique-based decompositions correspond to chordal graphs, each component of the mixture will be a junction tree, and $k^m(\mathbf{x})$ will be a probability distribution. In fact, a mixture of Kikuchi approximations opens the possibility of combining components that are probability distributions with other that are not.

To learn a mixture of Kikuchi approximations, a version of the EM algorithm was proposed. The general scheme is similar to the procedure used to learn mixture of trees [8]. However, fundamental differences arise in the expectation and maximisation steps. The sampling step used by MKA-EDA is based on the application of Gibbs sampling to each component of the mixture.

5.7.2 Combining Kikuchi Approximations with Inference Methods

Clique-based decomposition and other region-based decompositions can be also combined with belief-propagation algorithms for optimisation. This is done by the application of message-passing algorithms [13, 21], to obtain the most probable global state of a graphical model that assigns probabilities to the solutions according to its function value.

In [4], a new optimisation algorithm based on the use of Kikuchi approximations, a maximum variant of GBP, and a method for finding the most probable configurations was introduced. It is shown that the search for optimal solutions can be inserted in the dynamics of the GBP algorithm. Furthermore, it is proved that in order to identify the best solutions found during GBP, it is not necessary to evaluate the configurations. The Kikuchi approximation can be used to evaluate the function instead. The algorithm was tested using the Ising model as the benchmark problem.

The use of the most probable configurations appears as an alternative to the costly application of Gibbs sampling for invalid factorisations. The results presented in [4] and other works dealing with this type of sampling method [9, 19] point to the convenience of further investigating this type of sampling methods for undirected graphs. Other alternative ways for improving the efficiency of sampling methods for EDAs based on Markov networks is to use non local update Markov chain Monte Carlo methods that change the value of a block of variables instead of updating only one site of the vector [18]. The use of blocked Gibbs sampling can be incorporated into MN-EDA to accelerate the convergence of the sampling algorithm.

5.7.3 Generalised EDAs Based on Undirected Models

In [17], a generalised EDA is proposed providing flexibility to use models within MN-EDA and MN-FDA. The pseudocode of generalised EDA is shown in Figure 5.6.

The generalised EDA allows the use of different classes of graphical models at each generation. The model choice should be related to the complexity of the data and to the patterns of interaction between the components of the problem. In situations in which there are few interactions between the variables, we could choose a simple class of models and avoid more complex learning algorithms (e.g. those required by Bayesian networks). Choosing a simpler model can thus lead to an advantage in terms of computational time. Additionally the marginal probabilities of a probabilistic model with lower order dependencies could be more accurately estimated from small data samples.

Using different classes of graphical models during the search will also allow to incorporate different sampling techniques that determine different ways of searching for solutions. Therefore, the dynamic change of the probabilistic model will need an automatic procedure to select among the different types of graphical models.

Generalised EDA

1. Set $t \Leftarrow 0$. Generate M points randomly.
2. **do**
3. Select a set S of $N \leq M$ points according to a selection method.
4. Learn an undirected-graph-based representation of the dependencies in S.
5. Using the graph, determine a class of graphical models or approximation strategy to approximate the distribution of points in S.
6. Determine an inference algorithm to be applied in the graphical model.
7. Generate M new points from the model using the inference method.
8. $t \Leftarrow t+1$
9. **until** Termination criteria are met.

Fig. 5.6 Generalised estimation of distribution algorithm

Table 5.1 Approximation strategies, graphical models, and inference methods to be employed by EDAs based on undirected graphs.

Graphs	Graphical models	Inference
exact graph	univariate	PLS,MPC
	junction tree	PLS,MPC-BP
	junction graph	PLS,MPC-BP
	clique-based Kikuchi approx.	GS
	Bethe approximation	MPC-loopy BP
	Kikuchi approximation	MPC-generalised BP
subgraph	univariate	PLS,MPC
	junction tree	PLS,MPC-BP
	junction graph	PLS,MPC-BP
	clique-based Kikuchi approx.	GS
	Bethe approximation	MPC-loopy BP
	Kikuchi approximation	MPC-generalised BP
triangulated graph	junction tree	PLS,MPC-BP

Table 5.1 shows a number of alternatives for selecting a probability model according to the graph structure. Column 1 (Graphs) describes whether the approximation comprises all and only those dependencies in the graph (exact), a subgroup of the dependencies (subgraph) or all the dependencies and additional dependencies (triangulated graph). Column 2 (Graphical models) describes different situations that could be faced (e.g. univariate –there are not dependencies–, junction tree –valid factorisation–, etc.). Column 3 (Inference) shows different sampling algorithms that can be used according to the model. They comprise: probabilistic logic sampling (PLS), Gibbs sampler (GS), most probable configurations (MPC), most probable configurations with belief propagation (MPC-BP), most probable configurations with loopy belief propagation (MPC-loopy BP), and most probable configurations with generalised belief propagation (MPC-generalised BP).

5.8 Conclusions

Factorisations play a central role in EDAs. While the graphical models framework offers an impressive flexibility for modelling and solving optimisation problems, more work is required to satisfy all the goals involved in the optimisation, e.g. generality, accuracy, efficiency. In this chapter we have presented MN-EDA as a particular approach to the construction of factorisations in EDAs. MN-EDA is a general algorithm that has been conceived to solve problems for which no information about interactions between their components is available. However, when a priori information of the relationships between the variables of the problem is given, MN-EDA can use this information. Since the clique-based decompositions used by MN-EDA cover valid factorisations such as those represented by junction trees, the algorithm can be very accurate for problems with a simple factorised structure. The efficiency of the algorithm can be further enhanced by adapting its parameters according to the characteristics of the problems.

The chapter has also reviewed a number of extensions to MN-EDA. These developments pursue to increase the applicability and flexibility of the algorithm. Previous work has shown that mixtures of distributions, adaptive model class selection, and application of message-passing based inference methods can improve the results of MN-EDA. These lines of research are also promising for their application in the context of other EDAs.

Acknowledgements. This work has been partially supported by the Saiotek and Research Groups 2007-2012 (IT-242-07) programs (Basque Government), TIN2010-14931 and Consolider Ingenio 2010 - CSD 2007 - 00018 projects (Spanish Ministry of Science and Innovation).

References

1. Bron, C., Kerbosch, J.: Algorithm 457—finding all cliques of an undirected graph. Communications of the ACM 16(6), 575–577 (1973)
2. Geman, S., Geman, D.: Stochastic relaxation, Gibbs distributions, and Bayesian restoration of images. IEEE Transactions on Pattern Analysis and Machine Intelligence (6), 721–741 (1984)
3. Höns, R.: Estimation of Distribution Algorithms and Minimum Relative Entropy. PhD thesis, University of Bonn, Bonn, Germany (2006)
4. Höns, R., Santana, R., Larrañaga, P., Lozano, J.A.: Optimization by max-propagation using Kikuchi approximations. Technical Report EHU-KZAA-IK-2/07, Department of Computer Science and Artificial Intelligence, University of the Basque Country (November 2007)
5. Jakulin, A., Rish, I.: Bayesian Learning of Markov Network Structure. In: Fürnkranz, J., Scheffer, T., Spiliopoulou, M. (eds.) ECML 2006. LNCS (LNAI), vol. 4212, pp. 198–209. Springer, Heidelberg (2006)
6. Kikuchi, R.: A theory of cooperative phenomena. Physical Review 81(6), 988–1003 (1951)

7. Kirkpatrick, S., Gelatt, C.D.J., Vecchi, M.P.: Optimization by simulated annealing. Science 220, 671–680 (1983)
8. Meila, M., Jordan, M.I.: Learning with mixtures of trees. Journal of Machine Learning Research 1, 1–48 (2000)
9. Mendiburu, A., Santana, R., Lozano, J.A.: Introducing belief propagation in estimation of distribution algorithms: A parallel framework. Technical Report EHU-KAT-IK-11/07, Department of Computer Science and Artificial Intelligence, University of the Basque Country (October 2007)
10. Morita, T.: Formal structure of the cluster variation method. Progress of Theoretical Physics Supplements 115, 27–39 (1994)
11. Mühlenbein, H., Mahnig, T., Ochoa, A.: Schemata, distributions and graphical models in evolutionary optimization. Journal of Heuristics 5(2), 213–247 (1999)
12. Pakzad, P., Anantharam, V.: Belief propagation and statistical physics. In: Electronic Proceedings of 2002, Conference on Information Sciences and Systems. Princeton University, Paper No.225, CD-ROM, 3 pages (2002)
13. Pearl, J.: Probabilistic Reasoning in Intelligent Systems: Networks of Plausible Inference. Morgan Kaufmann, San Mateo (1988)
14. Santana, R.: A Markov Network Based Factorized Distribution Algorithm for Optimization. In: Lavrač, N., Gamberger, D., Todorovski, L., Blockeel, H. (eds.) ECML 2003. LNCS (LNAI), vol. 2837, pp. 337–348. Springer, Heidelberg (2003)
15. Santana, R.: Estimation of distribution algorithms with Kikuchi approximations. Evolutionary Computation 13(1), 67–97 (2005)
16. Santana, R., Larrañaga, P., Lozano, J.A.: Mixtures of Kikuchi Approximations. In: Fürnkranz, J., Scheffer, T., Spiliopoulou, M. (eds.) ECML 2006. LNCS (LNAI), vol. 4212, pp. 365–376. Springer, Heidelberg (2006)
17. Santana, R., Larrañaga, P., Lozano, J.A.: Adaptive estimation of distribution algorithms. In: Cotta, C., Sevaux, M., Sörensen, K. (eds.) Adaptive and Multilevel Metaheuristics. SCI, vol. 136, pp. 177–197. Springer (2008)
18. Santana, R., Mühlenbein, H.: Blocked stochastic sampling versus Estimation of Distribution Algorithms. In: Proceedings of the 2002 Congress on Evolutionary Computation CEC 2002, vol. 2, pp. 1390–1395. IEEE Press (2002)
19. Soto, M.R.: A Single Connected Factorized Distribution Algorithm and its Cost of Evaluation. PhD thesis, University of Havana, Havana, Cuba (July 2003) (in Spanish)
20. Spirtes, P., Glymour, C., Scheines, R.: Causation, Prediction and Search. Lecture Notes in Statistics, vol. 81. Springer, New York (1993)
21. Yanover, C., Weiss, Y.: Approximate inference and protein-folding. In: Becker, S., Thrun, S., Obermayer, K. (eds.) Advances in Neural Information Processing Systems 15, pp. 1457–1464. MIT Press, Cambridge (2003)
22. Yedidia, J.S., Freeman, W.T., Weiss, Y.: Understanding belief propagation and its generalizations. Technical Report TR-2001-22, Mitsubishi Electric Research Laboratories (November 2001)
23. Yedidia, J.S., Freeman, W.T., Weiss, Y.: Constructing free energy approximations and generalized belief propagation algorithms. Technical Report TR-2002-35, Mitsubishi Electric Research Laboratories (August 2002)

Part II
Theory

Chapter 6
Convergence Theorems of Estimation of Distribution Algorithms

Heinz Mühlenbein

Abstract. Estimation of Distribution Algorithms (EDAs) have been proposed as an extension of genetic algorithms. We assume that the function to be optimized is additively decomposed (ADF). The interaction graph of the ADF function is used to create exact or approximate factorizations of the Boltzmann distribution. Convergence of the algorithm MN-GIBBS is proven. MN-GIBBS uses a Markov network easily derived from the ADF and Gibbs sampling. We discuss different variants of Gibbs sampling. We show that a good approximation of the true distribution is not necessary, it suffices to use a factorization where the global optima have a large enough probability. This explains the success of EDAs in practical applications using Bayesian networks.

6.1 Introduction

The *Estimation of Distribution Algorithm* (EDA) family of population based optimization algorithms was introduced by [24] as an extension of genetic algorithms. They address the problem that the search distributions implicitly generated by genetic algorithms through recombination and crossover do not exploit the correlation of the variables in samples of high fitness values. Therefore genetic algorithms have difficulties in solving these problems.

EDAs use probability distributions derived from the function to be optimized to generate search points instead of crossover and mutation as done by genetic algorithms. The other parts of the algorithms are identical. In both cases a population of points is used and points with good fitness are selected either to estimate a search distribution or to be used for crossover and mutation. Today two major branches of EDAs can be distinguished. In the first branch a factorization of the distribution is computed from the mathematical expression of the function to be optimized, in

Heinz Mühlenbein
Fraunhofer Institut IAIS, Sankt Augustin, Germany

S. Shakya and R. Santana (Eds.): Markov Networks in Evolutionary Computation, ALO 14, pp. 91–108.
springerlink.com

the second one the factorization is computed from the probabilistic dependencies of the variables in samples of points with high fitness (learning). In both branches Bayesian networks are used mainly, especially if the network structure is learned. Often Bayesian networks cannot express all dependencies of the variables. In this paper we analyze EDAs using Markov networks. Several convergence theorems are proven and a short comparison to Bayesian networks is made.

The outline of the paper is as follows. In section 6.2 we introduce the Boltzmann distribution. Then factorizations of the distribution are discussed. For additively decomposed functions (ADFs) a factor graph [12] can easily be computed. The factor graph defines a Markov network. The algorithm MN-GIBBS is investigated which uses *Gibbs sampling* to generate samples of the distribution. A general convergence theorem is proven for MN-GIBBS.

In section 6.3 convergence results are presented for finite populations. We derive upper bounds on the sample size needed for a good approximation of the true distribution. The bounds are obtained by using the theory developed in *Probably Approximately Correct* (PAC) learning [11]. In section 6.3.1 we describe some algorithms which learn a Markov network from a sample of points. In section 6.4 we compare EDAs using Markov networks with EDAs using Bayesian networks. For Bayesian networks a simple sampling method exists. But only for a subset of Bayesian networks convergence to the optimum can be proven. Because the computational complexity of EDAs using Gibbs sampling is large compared to EDAs using Bayesian networks early implementations made simplifications concerning the Markov network or the sampling method. We propose several techniques which reduce the computational complexity of Gibbs sampling.

In section 6.5 we present numerical examples for large spin glass problems. The paper closes with a simple example which shows that exact factorizations of the distribution are not necessary to find the optimum. This partly explains the success of EDAs using Bayesian networks.

6.2 Convergence Theory for Infinite Samples

We will use in this paper the following notation. Capital letters denote variables, lower cases denote instances of variables. If the distinction between variables and instances is not necessary, we will use lower case letters. Vectors are denoted by $\mathbf{x}$, a single variable by x_i. We consider discrete variables only.
Let a function $f : \mathscr{X} \to \mathbb{R}$ be given. We consider the discrete optimization problem

$$\mathbf{x}_{opt} = argmax(f(\mathbf{x})) \tag{6.1}$$

A good candidate for optimization using a search distribution is the Boltzmann distribution.

Definition 1. *For $\beta \geq 0$ the* Boltzmann distribution[1] *of a function $f(\mathbf{x})$ is defined as*

$$p_\beta(\mathbf{x}) := \frac{e^{\beta f(\mathbf{x})}}{\sum_{\mathbf{y}} e^{\beta f(\mathbf{y})}} =: \frac{e^{\beta f(\mathbf{x})}}{Z_f(\beta)} \tag{6.2}$$

where $Z_f(\beta)$ is the partition function.

The Boltzmann distribution concentrates with increasing β around the global optima of the function. Obviously, the distribution converges for $\beta \to \infty$ to a distribution where only the optima have a probability greater than 0. Therefore, if it were possible to sample efficiently from this distribution for arbitrary β, optimization would be an easy task. But the computation of the partition function usually needs an exponential effort for a problem of n variables. We have therefore proposed an algorithm which incrementally computes the Boltzmann distribution from empirical data using Boltzmann selection.

Definition 2. *Given a distribution p and a selection parameter $\Delta\beta$,* Boltzmann selection *calculates the distribution for selecting points according to*

$$p^s(\mathbf{x}) = \frac{p(\mathbf{x})e^{\Delta\beta f(\mathbf{x})}}{\sum_y p(\mathbf{y})e^{\Delta\beta f(\mathbf{y})}} \tag{6.3}$$

The following theorem has been proven by [22].

Theorem 1. *If $p_\beta(\mathbf{x})$ is a Boltzmann distribution, then $p^s(\mathbf{x})$ is a Boltzmann distribution with inverse temperature $\beta(t+1) = \beta(t) + \Delta\beta(t)$.*

The following algorithm is called *BEDA*, the Boltzmann Estimated Distribution Algorithm.

BEDA

- **STEP 0:** $t \Leftarrow 0$. Generate N points according to the uniform distribution ($\beta(0) = 0$).
- **STEP 1:** With $\Delta\beta(g) > 0$ do Boltzmann selection giving the distribution $p^s(\mathbf{x}^t)$.
- **STEP 2:** Generate N new points according to the distribution $p(\mathbf{x}^{t+1}) = p^s(\mathbf{x}^t)$.
- **STEP 3:** If termination criteria fulfilled, STOP.
- **STEP 4:** $t \Leftarrow t+1$. GOTO **STEP 1**.

The following convergence theorem has been proven by [23].

Theorem 2. *For $\sum_t \Delta\beta(t) \to \infty$ and infinite populations BEDA converges to a distribution where only the global optima have a probability greater than zero.*

[1] The Boltzmann distribution is usually defined as $e^{-\frac{E(\mathbf{x})}{T}}/Z$. The term $E(x)$ is called the energy and $T = 1/\beta$ the temperature. We use the inverse temperature β instead of the temperature.

[16] have derived an adaptive annealing schedule SDS (standard deviation schedule) for Boltzmann selection and analyzed theoretically. It computes $\Delta\beta(t) > 0$ from the standard deviation of the fitness of the population. *BEDA* is only a conceptional algorithm, because the calculation of the distribution $p^s(\mathbf{x}^t)$ requires a sum over exponentially many terms. If the fitness function is additively decomposed the distribution can be factorized, either into a Bayesian network or more generally into a Markov network. These factorizations can be used to compute the distribution efficiently.

Definition 3. *Let $S_1,\ldots,S_m$, $S_i \subseteq \{1,\ldots,n\}$ be index sets. Let f_i be functions depending only on the variables of S_i. We denote these variables $\mathbf{x}_{s_i}$. S_i is called the scope of f_i. Then*

$$f(\mathbf{x}) = \sum_{i=1}^{m} f_i(\mathbf{x}_{s_i}) \tag{6.4}$$

is an additive decomposition *of the fitness function (ADF).*

Definition 4. *Given an ADF, the interaction graph G_{ADF} is defined as follows: The vertices represent the variables of the ADF. Two vertices are connected by an edge iff the corresponding variables are contained in the same sub-function.*

Remark: The class ADF covers all possible functions. In order to obtain algorithms of polynomial complexity we will later restrict the class. We will assume that the size of the scopes is bounded by a constant independently of n.

6.2.1 A Convergence Theorem for MN-GIBBS

In this section we investigate how to efficiently compute and sample the Boltzmann distribution for ADFs. The idea is to factorize the distribution into a product of factors. Bayesian networks will be discussed in section 6.4. But the most natural graphical representations of the structure of ADFs are *factor graphs* [12].

Definition 5. *A factor graph $\mathscr{FG}$ is a graph with two kinds of vertices, the set of factors $\{F_j\}_{j=1}^m$ with scopes $\{S_j\}_{j=1}^m$, and the set of random variables $\mathscr{X} = \{X_1,\ldots,X_n\}$. Each variable is connected to those factors where it is contained in the scope. The Gibbs distribution of $\mathscr{FG}$ is defined as*

$$p_G(\mathbf{x}) = \frac{1}{Z}\prod_{j=1}^{m} F_j(\mathbf{x}_{s_j}) \tag{6.5}$$

The Gibbs distribution defines a Markov network or Markov random field *on the variables [26].*

The Boltzmann distribution of an ADF can easily be written as a Gibbs distribution.

$$p_\beta(\mathbf{x}) = \frac{1}{Z_f(\beta)} e^{\beta \sum_{j=1}^m f_j(\mathbf{x}_{s_j})} = \frac{1}{Z_f(\beta)} \prod_{i=1}^{m} e^{\beta f_j(\mathbf{x}_{s_j})} \tag{6.6}$$

Now set $F_j(\mathbf{x}_{s_j}) = e^{\beta f_j(\mathbf{x}_{s_j})}$ and a Gibbs distribution is obtained which is identical to the Boltzmann distribution. Therefore the BEDA convergence theorem remains valid for factor graphs. Because the computation of Z is exponential in n, the above factorization is no improvement at first sight. But sampling from a Gibbs distribution can be done using local computations only. The method is called *random Gibbs sampling* [5, 26][2]. In order to understand Gibbs sampling, the concept *Markov blanket* is needed, the minimal set of variables that separate all variables of a given set D from the other variables in the graph.

Definition 6 (Markov Blanket ([26])). *Let a set of scopes S_j be given. The Markov blanket of a set of variables $D \subseteq \mathscr{X}$ is defined as*

$$MB(D) = \bigcup \{S_j : S_j \in S, S_j \cap D \neq \emptyset\} \setminus D \tag{6.7}$$

Random Gibbs Sampling

- **STEP 0:** Generate $\mathbf{x}^0 = (x_1^0, \ldots, x_n^0)$ randomly. Set $t = 0$.
- **STEP 1:** Choose an index i randomly.
- **STEP 2:** Sample x_i^{t+1} using $p(x_i|MB(X_i))$
- **STEP 3:** Set $\mathbf{x}^{t+1} = (x_1^t, \ldots, x_i^{t+1}, \ldots x_n^t)$.
- **STEP 4:** t:= t+1; If $t \leq Max$ goto **STEP 1**.

Example 1:

$$f(\mathbf{x}) = f_1(x_1, x_2) + f_2(x_2, x_3) + f_3(x_3, x_4) + f_4(x_4, x_1) \tag{6.8}$$

The dependencies of G_{ADF} form a loop. We have

$$S_1 = \{X_1, X_2\}, S_2 = \{X_2, X_3\}, S_3 = \{X_3, X_4\}, S_4 = \{X_4, X_1\}$$

Sequential Gibbs sampling updates the variables as follows

$$p(x_1^t | x_2^{t-1}, x_4^{t-1}), p(x_2^t | x_1^t, x_3^{t-1}), p(x_3^t | x_2^t, x_4^{t-1}), p(x_4^t | x_3^t, x_1^t), \ldots$$

Gibbs sampling uses only local computations. If the size of the Markov blankets is bounded polynomially, the computational complexity of one iteration is polynomially bounded. The first samples (until $t = t_0$) are usually thrown away.

Instead of updating a single variable only, it is possible to update a set of variables. This is called *blocked Gibbs sampling*. Given a set D the Gibbs sampling formula has to be changed in STEP 2:

- **STEP 2:** Sample $\mathbf{x}_D^{t+1}$ using $p(\mathbf{x}_D|MB(D))$

For the above example blocked Gibbs sampling can update two variables in a step

$$p(x_1^t, x_2^t | x_3^{t-1}, x_4^{t-1}), p(x_3^t, x_4^t | x_1^t, x_2^t), \ldots$$

[2] [26] calls it stochastic simulation.

Blocked Gibbs sampling converges faster because it updates connected variables in one step. This is also true in the context of EDAs as was shown by [29].
We now prove that Gibbs sampling converges to the true distribution.

Theorem 3 (Convergence). *Let $\mathscr{FG}$ be the factor graph corresponding to the given ADF. Then Gibbs sampling converges to the true Gibbs distribution if all conditionals are greater than zero.*

Proof: The proof is based on two basic theorems in the field of Markov chains and Markov random fields. We recall

Definition 7. *The stochastic process $X^{(i)} = (x_1, \dots, x_n)$ is called a Markov chain if*

$$T(X^{(i)}|X^{(i-1)}, \dots, X^{(0)}) = T(X^{(i)}|X^{(i-1)}) \tag{6.9}$$

T is called the transition matrix. The transition to $X^{(i)}$ depends only on the states of $X^{(i-1)}$. The interested reader is referred to [3] and [5, 7]. Obviously Gibbs sampling defines a Markov chain. The transition matrix is given by the products of the local transitions used for Gibbs sampling. The following theorem is the foundation of Markov chains [3].

Theorem 4. *A Markov chain converges to a stationary distribution $\pi^*(x)$ if the chain has the two properties*

1. *Irreducibility*
2. *Aperiodicity*

Irreducibility means that there is a positive probability of visiting all states. Gibbs sampling fulfills the assumptions of the above theorem if $p(\mathbf{x}) > 0$ for all $\mathbf{x}$. Furthermore Gibbs sampling is aperiodic.

We now come to the difficult part of the proof. We know that Gibbs sampling converges to a stationary distribution, but not the structure of this distribution. This problem is solved by the next theorem. We state the theorem in the notation of [5].

Theorem 5. *Let $\mathscr{G}$ be the neighborhood system defined by the factor graph of the ADF. Perform Gibbs sampling using this neighborhood. Then X is a Markov random field with respect to $\mathscr{G}$ if and only if the stationary distribution $\pi^*(\mathbf{x})$ is the Gibbs distribution with respect to $\mathscr{G}$.*

In our application the Markov random field is defined by the factors and the corresponding Gibbs distribution. Therefore the stationary distribution of Gibbs sampling is this Gibbs distribution.

We can now define the Markov network algorithm (**MN-GIBBS**).

MN-GIBBS

- **STEP 0:** Compute the Markov network from the factor graph of the ADF.
- **STEP 1:** $g \Leftarrow 0$. Generate N points according to the uniform distribution with $\beta(0) = 0$.

- **STEP 2:** With a given $\Delta\beta(t) > 0$ do Boltzmann selection.
- **STEP 3:** Compute the conditional probabilities $p(x_i|MB(X_i))$ using the selected points.
- **STEP 4:** Generate a new population N^g according to Gibbs sampling.
- **STEP 5:** If termination criteria are met, STOP.
- **STEP 6:** Add the best point of the previous generation to the generated points (elitist).
- **STEP 6:** Set $g \Leftarrow g+1$. Goto **STEP 2.**

From theorems 2 and 5 follows the next theorem.

Theorem 6. *For $\sum_g \Delta\beta(g) \to \infty$ and infinite populations the algorithm MN-GIBBS converges to a distribution where only the global optima have a probability greater than zero.*

The computational complexity for updating n variables is bounded by $O(n)$ if the scope of the Markov blanket is bounded by a constant independent of n. But convergence to the true distribution has been proven only for an infinite number of Gibbs samples. In the next section we will bound the population size N, but a bound on the number of Gibbs samples can only be done for special cases.

We want to conclude this section with a remark by [15] (p.397). "An ideal genetic algorithm would be one that can be proved to be a valid Monte Carlo algorithm that converges to a specified density." MN-GIBBS is such an algorithm.

6.2.2 Some Implementations of EDAs Using Markov Networks

An early implementation using a restricted class of Markov networks is reported in [27]. The algorithm called *MN-EDA^f* by [28] is an implementation of MN-GIBBS. The author is mainly interested in learning the Markov network from data, so *MN-EDA^f* is not thoroughly investigated. In the paper also different variants of Gibbs sampling are numerically investigated. An interesting variant is to start sampling not randomly, but at points with high fitness values.

The algorithm IS-DEUM by [30] is a substantially modified Markov network algorithm. The author does not use the ADF for constructing the Markov network, but computes a restricted model U of the fitness function from the factor equation $-\ln f(\mathbf{x}) = U(\mathbf{x})$ using N samples. The author considers linear and quadratic U. [28] and [18] introduced *region graphs* for EDAs. Region graphs are closely related to Markov random fields and the Gibbs distribution. Both authors use the Kikuchi approximation to the region graphs.

$MARLEDA^{+model}$ from [2] seems to be a full implementation of MN-GIBBS. The Markov network model is derived from the ADF. Gibbs sampling is used to create a new population. The author does not investigate this algorithm, because he concentrates on the difficult task of learning the Markov model from data.

6.3 Convergence of MN-GIBBS with Finite Samples

The convergence theorems presented so far are valid for infinite populations only. We now investigate convergence for finite populations. EDAs are probabilistic algorithms. For finite populations convergence to the global optima can only be probabilistic. A concept which covers this aspect is (ε, δ) convergence. It was first applied in *Probably Approximately Correct* (PAC) learning [11]. It is defined as follows:

Definition 8. *Let* $\varepsilon > 0,\ \delta > 0$. *Let p be the true distribution,* $\tilde{p}$ *an approximation. Then we speak of* (ε, δ) *convergence if*

$$prob(error(p, \tilde{p}) \leq \varepsilon) \geq 1 - \delta \tag{6.10}$$

error denotes any distance measure.
Let p be a probability distribution over n binary variables. Let

$$D_N = \{\mathbf{X}^1, \ldots, \mathbf{X}^N\}, \mathbf{X}^i = (x_1^i, \ldots, x_n^i)$$

be samples drawn according to p. It is easy to show that the maximum likelihood approximation of the true probabilities are the long-run frequencies of the sample. This gives the approximation q. The error between two distributions is often measured by the Kullback-Leibler divergence.

Definition 9. *The Kullback-Leibler (KLD) divergence between two distributions is defined by*

$$D(p||q) = \sum_{\mathbf{x}} p(\mathbf{x}) \ln \frac{p(\mathbf{x})}{q(\mathbf{x})} \tag{6.11}$$

Note that the divergence is not symmetric!

The sample size complexity has been estimated by [1] (theorem 6). Let k be the maximum number of variables in a scope C_j. Let b be the maximum number of variables in Markov blanket. Let

$$\gamma = min_{X_i} p(X_i = x_i | MB(X_i)).$$

Theorem 7. *Let any* ε, δ *be given. Let be given (a) N training examples drawn i.i.d. from a distribution p and (b) the factor graph according to which the distribution p factors. Let* $\tilde{p}$ *be the probability distribution with the factors computed empirically. Then, we have that, for*

$$prob(KLD(p, \tilde{p}) < \varepsilon) > 1 - \delta \tag{6.12}$$

to hold that the number of training examples N satisfies

$$N \geq \left(1 + \frac{\varepsilon}{2^{2k+2}}\right)^2 \frac{2^{4k+3}}{\gamma^{2k+2b} \varepsilon^2} m \log \frac{2^{2k+b+2} m}{\delta} \tag{6.13}$$

If k,b and γ are bounded independently from n then the sample size should be $N \in O(n \log n)$. This is astonishingly small. Thus estimating the conditional probabilities of the Markov network is not a problem for MN-GIBBS.

The theorem can be used to obtain an (ε, δ) convergence theorem for MN-GIBBS. Here convergence means the probability to generate the global optima. The proof is difficult, it assumes that Gibbs sampling has converged. Therefore the theorem is of limited practical value, as the computational complexity of MN-GIBBS is dominated by Gibbs sampling.

6.3.1 Learning of Markov Networks and Factor Graphs

Learning of factor graphs is easier than learning of Bayesian networks because they match the ADF structure. [1] present a polynomial learning algorithm. The method uses the conditional entropy to find the best network in a restricted class of Markov networks, where k and b are fixed, independent of n. The computational complexity cc of this learning method is given by

$$cc \in O\left(Nkb(k+b)n^{k+b}\right) \tag{6.14}$$

The learning algorithm tests all combinations of sets of variables and corresponding Markov blankets. Therefore we have an exponential dependence on the maximum scope size k and the maximum Markov blanket size b, the dominating term is n^{k+b}. For large n the algorithm is already computationally too expensive for $k \geq 3$. The polynomial bound can be made smaller by implementing a more sophisticated learning algorithm. To my knowledge the learning algorithm has not yet been implemented. The sample complexity for a good approximation of the true distribution remains small. It suffices that $N \in O(n \log n)$ [1].

A much simpler learning algorithm is used in MARLEDA [2]. It applies Pearson's χ^2 test to compute the confidence level of the dependencies between variables. Using the dependencies a Markov network is constructed. The construction procedure is very simple, so it does not guarantee that a correct factor graph is obtained. MN-EDA [28] also uses the χ^2 test to detect dependent variables. It creates not a full Markov network, but Kikuchi approximations.

6.3.2 Sample Complexity for Learning the ADF Structure by Probing

A learning method not based on statistical independence tests has been investigated by [9]. Earlier work has been reported by [25]. The method computes the ADF structure of the function by computing its Walsh coefficients.

Theorem 8 ([9]). *Assume a class of ADF functions where each sub-function has at most k variables. Let $\delta > 0$ be a constant. Then the number of function evaluations required by the proposed learning algorithm of order 2 to detect the scope of all sub-functions with probability at least $1 - \delta$ is bounded by*

$$N \in O\left(2^k n^2 \ln n \ln(1 - \delta^{\frac{1}{2}})\right) \tag{6.15}$$

This learning algorithm has a computational complexity of only $O(n^2 \ln n)$. It is worthwhile to be explored in EDAs. The learned ADF structure can be used to create a graphical model. [31] propose to use the Factorized Distribution Algorithm FDA to compute a Bayesian factorization of the ADF. FDA is explained next.

6.4 A Comparison of Markov Networks and Bayesian Networks

The structure of the graphical model of the ADF has to be restricted in order that sampling becomes easy. The most popular models are Bayesian networks. Our Factorized Distribution Algorithm FDA computes a Bayesian network from the ADF which is without loops. The construction of the factorization is described in more detail in ([19, 20]).

Definition 10. *Given the scopes of the ADF $S_1, \ldots, S_m$, we define the sets D_i, B_i and C_i for $i = 1, \ldots, m$:*

$$D_i := \bigcup_{j=1}^{i} S_j, \tag{6.16}$$

$$B_i := S_i \setminus D_{i-1}, \tag{6.17}$$

$$C_i := S_i \cap D_{i-1} \tag{6.18}$$

We require that $(B_i \neq \emptyset,\ p(\mathbf{x}_{B_i}|\mathbf{x}_{C_i}) > 0\ : i = 1, \ldots, m)$, $D_m = \{1, \ldots, n\}$ and set $D_0 = \emptyset$. A FDA factorization of the ADF is defined by

$$\tilde{p}(\mathbf{x}) = \prod_{i=1}^{m} p(\mathbf{x}_{B_i}|\mathbf{x}_{C_i}) \tag{6.19}$$

In the theory of decomposable graphs, D_i are called histories, B_i residuals *and C_i* separators *[13]. The set $\{B_i, C_i\}$ is a* clique[3].

The FDA factorization can easily be transformed into a Bayesian network. One has to use repeatedly

$$p(x_1, x_2, x_3, \ldots) = p(x_1) p(x_2|x_1) p(x_3|x_1, x_2) \ldots$$

[3] A clique [26] is a set of vertices V such that for every two vertices in V, there exists an edge connecting it. This is equivalent to saying that the subgraph induced by V is a complete graph.

Any FDA factorization can be used for sampling. The simplest sampling method is called *probabilistic logic sampling* (PLS) introduced by [10]. It works as follows:

Probabilistic Logic Sampling

- **STEP 1:** For $t = 1$ to N; For $i = 1$ to m
- **STEP 2:** Sample $\mathbf{x}^t_{B_i}$ from $p(\mathbf{x}_{B_i}|\mathbf{x}^t_{C_i})$

It is often overlooked that PLS *will in general not reproduce the true distribution*! It needs a severe additional restriction, the RIP.

Definition 11. *The running intersection property RIP [13] is fulfilled if*

$$\forall i \geq 2 \ \exists j < i \ C_i \subseteq S_j \tag{6.20}$$

Let us discuss the RIP with a simple example.

Example 2:

$$f(\mathbf{x}) = f_1(x_1,x_2) + f_2(x_2,x_3) + f_3(x_3,x_4) + f_4(x_4,x_1) \tag{6.21}$$

The ADF network is a loop. Loops are not allowed in FDA factorizations. The FDA factorization algorithm will compute the factorization

$$p(\mathbf{x}) = p(x_1,x_2)p(x_2|x_1)p(x_3|x_2)p(x_4|x_3) \tag{6.22}$$

In the above factorization the edge between x_4 and x_1 is missing. Inserting this edge gives the factorization

$$p(\mathbf{x}) = p(x_1,x_2)p(x_2|x_1)p(x_3|x_2)p(x_4|x_1,x_3) \tag{6.23}$$

This factorization violates the RIP. Thus PLS will not generate the true distribution. The next factorization is correct

$$p(\mathbf{x}) = p(x_1,x_2)p(x_2|x_1)p(x_3|x_2,x_1)p(x_4|x_1,x_3) \tag{6.24}$$

There exists an algorithm, called the *junction tree algorithm*, which creates for any Bayesian network a Bayesian network which fulfills the RIP [13]. But the network fulfilling the RIP might contain cliques of size $O(n)$, thus the computation might be prohibitive [19]. Instead of computing a junction tree and then use PLS one can extend the sampling method to Bayesian networks with loops. This sampling method is called *loopy belief propagation* LBP. The drawback of this method is that convergence of LBP cannot be proven. And even if it converges, it might not converge to the true distribution. Nevertheless, the numerical results of LBP in machine learning are promising. [17] have used it quite succesfully in an EDA.

The computational complexity of FDA with PLS is given by $O(G \cdot N \cdot n \cdot k_{max})$, where G is the number of generations and k_{max} is the size of the largest clique. FDA has experimentally proven to be very successful on a number of functions where standard genetic algorithms fail to find the global optimum. For recent surveys and a more detailed description of the algorithm, the reader is referred to [18, 19, 21, 22].

6.4.1 A Sampling Method Finding the Most Probable States

[32] have proposed an interesting new sampling method for Bayesian networks with loops. The algorithm tries to find the M *most probable* configurations. If we recall that the most probable configurations are those with the highest fitness values, the algorithm is ideally suited for optimization. Very few samples should be sufficient.

It is outside the scope of this paper to describe the algorithm and its theoretical foundation. It is based on loopy belief propagation. The algorithm is very efficient. For a large 32*32 spin glass model the authors found that their algorithm outperformed Gibbs sampling by far in generating configurations with high probability given the same number of steps.

The spin glass model is described next. Both from theoretical and numerical aspects we highly recommend to test this algorithm at least on problems defined on grids.

6.4.2 Computational Complexity of Gibbs Sampling

The computational complexity of MN-GIBSS is dominated by Gibbs sampling. Gibbs sampling is an iterative process, the number of steps needed for convergence to the true distribution is unknown in general. Convergence is defined as follows. Let a Markov chain with probability transition matrix T and stationary distribution π be given. Let $\mathbf{x}^0$ be the initial configuration.

Definition 12. *The total variation distance at step t is*

$$D_{x^0}(t) = \frac{1}{2}\sum_x |T^t(\mathbf{x}^0,\mathbf{x}) - \pi(\mathbf{x})| \tag{6.25}$$

The convergence time of the Markov chain used by the Gibbs sampler is defined as

$$\tau(\varepsilon) = \max_{x^0} \min\{t : D_{x^0}(t) \leq \varepsilon\} \tag{6.26}$$

$T^t(\mathbf{x}^0,\mathbf{x})$ denotes the probability that the Markov chain with initial state $\mathbf{x}^0$ is in state $\mathbf{x}$ at iteration t.

A survey of convergence times and methods for their computation has been done by [14]. For special functions in statistical physics and image restoration the computational complexity of Gibbs sampling has been intensively studied. A specific problem, the spin glass model will be discussed in the next section.

Blocked Gibbs sampling converges faster than single variable update, but the convergence might still be very slow. In particular, samples with high probability $p(\mathbf{x})$ might be generated only after a large number of steps. This problem is addressed by *importance sampling*.

Efficient sampling of Markov networks is still an active research area. A recent survey of different variants of Gibbs sampling has been published by [8]. In general, Gibbs sampling faces two major problems

- The first configuration is randomly chosen. Therefore it takes some time that Gibbs sampling generates the samples according to the given distribution.
- If Gibbs sampling generates a configuration with high probability and the surrounding configurations have low probability, it will remain there for a long time.

In order to reduce these problems we recommend the following strategies for the implementation.

1. Use blocked Gibbs sampling, preferable using the factors of the factor graph.
2. Do not start randomly, but compute x^0 according to the probabilities of the Markov blankets $p(MB(D))$.
3. Use very short runs, maybe just selecting one point in each run.

6.5 Theoretical and Numerical Results for the Ising Problem

The Ising spin glass on a 2D gris is defined as follows

$$f_{Ising} = -\sum_{i,j} J_{i,j} s_i s_j - \sum_i h_i s_i \tag{6.27}$$

j are the four nearest neighbors of i in the 2D grid. The couplings $J_{i,j}$ are randomly drawn from a Gaussian distribution. h is the external magnetic field. The spins s_i have values in $\{-1,1\}$. This function is symmetric, therefore it has two global optima, which are the binary complement from each other.

The Markov network for the Ising model is straightforward. For single spin update the blanket size is at most 4. In contrast, any Bayesian network which fulfills the RIP contains cliques of size $O(\sqrt(n))$ [19]. Therefore one has to use Bayesian networks which violate the RIP. But then PLS will not reproduce the true distribution, thus convergence to the optimum cannot be guaranteed. Thus here we have a problem where EDAs with Bayesian networks have to use approximate factorizations, whereas the Markov network is exact. The Ising model has been extensively studied in physics. Let us review the most important results. The ferromagnetic model is obtained if for all i,j $J_{i,j} = \beta > 0$. For this model [6] reports the convergence rate of Gibbs sampling. The convergence rate for one-dimensional problems is $O(n\ln(n))$ where n denotes the number of grid points. In dimensions higher than one, this result holds for $h = 0$ for all values below a critical value β_c at which a phase transition occurs[4]. For the Ising model with an external field ($h > 0$) the convergence rate can be shown to be $O(n\ln(n))$ for all β in two dimensions and for small β in higher dimensions. The convergence rate of $O(n\ln(n))$ seems to be the best possible for any random optimization algorithm.

[4] For two dimensional problems we have $\beta_c \approx 0.44$.

A more challenging problem is the Ising spin glass model. For this problem the convergence rates of Gibbs sampling have not yet been estimated. But the following computational complexity result has been obtained by [4]. The Ising spin glass problem defined on a two-dimensional lattice can be solved in polynomial time of the size of the lattice if there is no external field. In contrast, the same problem belongs to the class of **NP-hard** problems, both in the two-dimensional case with an external field, and in the three-dimensional case. In table 1 we present numerical results of the Ising spin glass. For each problem size we investigated 10 randomly generated problems. The FDA factorization uses cliques of size 5 [19]. LFDA learns Bayesian networks with at most 4 parents from the data [20].

Table 1 Best(1) and worst results(2) out of 10 randomly generated spin glass problems; (*) global optimum; δ percentage of not finding the optimum

n	$Alg.$	$Pr.$	N	$gen.$	δ	average	best
100	FDA	1	1000	13.5	0.2	73.890	73.977(*)
100	LFDA	1	1500	14.1	0.1	73.947	73.977(*)
100	FDA	2	3500	14.5	0.2	73.342	73.359(*)
100	LFDA	2	3000	13.1	0.1	73.353	73.359(*)
225	FDA	1	3000	24.5	0.2	168.494	168.540(*)
225	LFDA	1	3000	25.0	1.0	167.449	167.757(-)
225	FDA	2	4000	23.5	1.0	164.441	164.441(-)
225	FDA	2	70000	22.5	1.0	164.441	164.441(-)
225	LFDA	2	4000	25.5	0.7	164.304	164.473(*)
400	FDA	1	5000	33.1	0.8	298.874	300.504(*)
400	FDA	1	10000	33.5	0.9	298.467	300.504(*)
400	LFDA	1	5000	35.9	0.9	297.872	300.504(*)
625	FDA	1	10000	52.1	1.0	463.445	464.793(-)
400	M^+	-	900	20.0	-	-	(*)

The table shows the results of two instances giving the best (1) and the worst results (2). We also report the average of the best solution found. For $n = 100$ there is no difference in the performance between the different problems. For $n = 225$ we generated a problem instance (2) where FDA always converged to a local optimum. The value of the local optimum differs from the global optimum at the fifth decimal place. Increasing the pop-size from $N = 4000$ to $N = 70000$ did not improve the results. Interestingly LFDA found the optimum in 3 out of 10 runs. The optimum for $n = 400$ is found only once or twice in ten runs. Increasing the population size does not improve the results. For $n = 625$ the optimum is not found in 20 runs. The best solution is 0.5% less than the optimum. This shows that for larger problems too many dependencies of the 2D grid are missing in the FDA factorization. We have included a result from $MARLEDA^{+model}$ [2], despite it is a single run for an Ising model restricted to integer couplings ($J_{ij} \in \{-1,+1\}$). The algorithm found the optimum. The size of the Markov blanket is at most 4. [28] also reports results for his MN-GIBBS implementation $MN\text{-}EDA^f$ for very small Ising problems (up to $n = 64$). He remarkss that *MN-EDA*^{f} is the best algorithm among the algorithms he compared. But the runtime is very large so that larger problem instances have not been tried.

It is worthwhile to investigate the variants of Gibbs sampling more intensively for the Spin glass problems. The algorithm should converge in polynomial time, at least for the two-dimensional Ising spin glass problem without magnetic field.

6.6 Exact and Approximate Factorizations of the True Distribution

In the paper we have discussed several graphical models and sampling methods. Let us now discuss which graphical model is best suited for a particular application. We have proven two theoretical convergence results. FDA will find the optimum if the factorization used contains all edges of G_{ADF} and fulfills the RIP. MN-GIBBS will always find the optimum, but is more costly.

Both theorems give *sufficient conditions*, they are not necessary for finding the optimum. In EDA applications we are interested to detect the global optima, and a good approximation of the true Boltzmann distribution is not necessary to achieve this goal. We discuss the problem with a simple example.

Example 3:

$$f(\mathbf{x}) = \sum_{i=1}^{n}(1 - x_i) + (n+1)\prod_{i=1}^{m} x_i \tag{6.28}$$

The exact factorization is

$$p(\mathbf{x}) = p(x_1, \ldots, x_m) \prod_{i=m+1}^{n} p(x_i) \tag{6.29}$$

If $m = n$ the optimization is a needle in a haystack problem. If m is bounded independently from n, FDA can solve the optimization problem in polynomial time using the above factorization. But the exact factorization is not necessary to find the optimum. For $m = 2$, for example, the simple factorization $p(\mathbf{x}) = \prod_{i=1}^{n} p(x_i)$ is sufficient if the population size N is large enough. If we change the function to

$$f(\mathbf{x}) = \sum_{i=1}^{n} x_i + (n+1)\prod_{i=1}^{m} x_i \tag{6.30}$$

the optimization becomes easy. The exact factorization is again given by (6.29). But FDA using the simple factorization $p(\mathbf{x}) = \prod_{i=1}^{n} p(x_i)$ will find the optimum for all m! Both functions have the same interaction graph. The second optimization problem is easy, whereas the computational complexity to optimize the first one ranges from polynomial to exponential if m is increased.

This leads us to the main conclusion: Instead of using the exact distribution, it is always possible to use a *simpler distribution which has the same global optima as the original distribution and selection increases the probabilities of the optima*. This explains partly the numerical success of EDAs using simple factorizations which do not produce the true distribution.

6.7 Conclusion and Outlook

We have presented theoretical arguments to use Markov networks for EDAs. If the Markov network is correctly derived from the function to be optimized, MN-GIBBS will find the optimum. But there exists a severe implementation problem. While the computation of the conditional probabilities and even learning of the Markov network structure can be efficiently done, efficient sampling of Markov networks is still an unsolved problem. An interesting alternative is using Bayesian networks with loops and loopy belief propagation. Promising is here an algorithm which generates the most probable configurations.

Optimization problems defined on grids are a good start to compare Markov networks with Bayesian networks. Bayesian networks with bounded clique size do not reproduce the true distribution. Nevertheless. for Ising spin glasses on 2D grids Bayesian networks with clique sizes of at most six give good numerical results.

Markov networks have blankets of size at most 4, but the runtime of a standard implementation of Gibbs sampling is large for problems of size greater than 400. We have proposed several methods for doing the sampling more efficient. Only if the sampling method used for the Markov network gives better results than EDAs using Bayesian networks in comparable time, Markov networks will be an alternative to Bayesian networks in EDAs.

References

1. Abbeel, P., Koller, D., Ng, A.: Learning factor graphs in polynomial time & sample complexity. Journ. Machine Learning Research 7, 1743–1780 (2006)
2. Alden, M.E.: MARLEDA: Effective Distribution Estimation through Random Fields. PhD thesis, University of Texas at Austin, Austin, USA (2007)
3. Andrieux, C., de Freitas, N., Doucet, A., Jordan, M.: An introduction to MCMC for machine learning. Machine Learning 50, 5–43 (2003)
4. Barahona, F.: On the computational complexity of the Ising spin glass models. J. Phys. A: Math. Gen. 15, 3241–3253 (1982)
5. Geman, D., Geman, S.: Stochastic relaxation, Gibbs sampling, and the restauration of images. IEEE Transaction on Pattern Recognition and Machine Intelligence 6, 721–741 (1984)
6. Gibbs, A.: Bounding the convergence time of the Gibbs sampler in Bayesian image restoration. Biometrika 87, 749–766 (2000)
7. Gilks, W.R., Richardson, S., Spiegelhalter, D.J.: Markov Chain Monte Carlo in Practice. Chapman & Hall, London (1996)
8. Guo, H., Hsu, W.: A survey of algorithms for real-time Bayesian network inference. In: KDD 2002/UAI 2002 Workshop on Real-Time Decision Support (2002), `http://citeseer.ist.psu.edu/Guo2survey.html`

9. Heckendorn, R.B., Wright, A.H.: Efficient linkage discovery by limited probing. Evolutionary Computation 12, 517–545 (2004)
10. Henrion, M.: Propagating uncertainty in Bayesian networks by Probabilistic Logic Sampling. In: Lemmar, J., Kanal, L. (eds.) Uncertainty in Artificial Intelligence, pp. 149–181. Elsevier, New York (1988)
11. Kearns, M., Vazirani, U.: An Introduction to Computational Learning Theory. MIT Press, Cambridge (1994)
12. Kschischang, F.R., Frey, B.J., Loeliger, H.A.: Factor graphs and the sum-product algorithm. IEEE Transactions on Information Theory 47(2), 498–519 (2001)
13. Lauritzen, S.L.: Graphical Models. Clarendon Press, Oxford (1996)
14. Levin, D.A., Peres, Y., Wilner, E.L.: Markov Chains and Mixing Times. American Mathematical Society, New York (2009)
15. MacKay, D.J.: Infomation Theory, Inference, and Learning Algorithms. Cambridge University Press, Cambridge (2003)
16. Mahnig, T., Mühlenbein, H.: A new adaptive Boltzmann selection schedule SDS. In: Proceedings of the Congress on Evolutionary Computation 2001, pp. 121–128. IEEE (2001)
17. Mendiburu, A., Santan, R., Lozano, J.: Introducing belief propagation in Estimation of Distribution Algorithms: A Parallel Framework. Technical Report EHU-KAT-IK-11-07, Intelligent Systems Group (2007)
18. Mühlenbein, H., Höns, R.: The estimation of distributions and the minimum relative entropy principle. Evolutionary Computation 13(1), 1–27 (2005)
19. Mühlenbein, H., Höns, R.: The factorized distribution algorithm and the minimum relative entropy principle. In: Pelikan, M., Sastry, K., Cantu-Paz, E. (eds.) Scalable Optimization via Probabilistic Modeling. SCI, pp. 11–37. Springer, Berlin (2006)
20. Mühlenbein, H., Mahnig, T.: FDA - a scalable evolutionary algorithm for the optimization of additively decomposed functions. Evolutionary Computation 7(4), 353–376 (1999)
21. Mühlenbein, H., Mahnig, T.: Evolutionary optimization and the estimation of search distributions with applications to graph bipartitioning. Journal of Approximate Reasoning 31(3), 157–192 (2002)
22. Mühlenbein, H., Mahnig, T.: Evolutionary algorithms and the Boltzmann distribution. In: DeJong, K., Poli, R., Rowe, J.C. (eds.) Foundations of Genetic Algorithms 7, pp. 525–556. Morgan Kaufmann Publishers, San Francisco (2003)
23. Mühlenbein, H., Mahnig, T., Ochoa, A.: Schemata, distributions and graphical models in evolutionary optimization. Journal of Heuristics 5(2), 213–247 (1999)
24. Mühlenbein, H., Paaß, G.: From Recombination of Genes to the Estimation of Distributions I. Binary Parameters. In: Voigt, H.-M., Ebeling, W., Rechenberg, I., Schwefel, H.-P. (eds.) PPSN 1996. LNCS, vol. 1141, pp. 178–187. Springer, Heidelberg (1996)
25. Munetomo, M., Goldberg, D.: Linkage identification by non-monotonicity detection for overlapping functions. Evolutionary Computation 7, 377–398 (1999)
26. Pearl, J.: Probabilistic Reasoning in Intelligent Systems. Morgan Kaufmann, San Mateo (1988)
27. Santana, R.: A Markov Network Based Factorized Distribution Algorithm for Optimization. In: Lavrač, N., Gamberger, D., Todorovski, L., Blockeel, H. (eds.) ECML 2003. LNCS (LNAI), vol. 2837, pp. 337–348. Springer, Heidelberg (2003)
28. Santana, R.: Estimation of distribution algorithms with Kikuchi approximations. Evolutionary Computation 13(1), 67–97 (2005)

29. Santana, R., Mühlenbein, H.: Blocked stochastic sampling versus Estimation of Distribution Algorithms. In: Proceedings of the Congress on Evolutionary Computation 2002, pp. 1390–1395. IEEE Press (2002)
30. Shakya, S.K.: DEUM: A framework for an Estimation of Distribution Algorithm based on Markov Random Fields. PhD thesis, Robert Gordon University, Aberdeen, Scotland (2006)
31. Wright, A.H., Pulavarty, S.S.: Estimation of distribution algorithm based on linkage discovery and factorization. In: Beyer, H.G., et al. (eds.) Proceedings of GECCO 2005, pp. 695–703. ACM Press (2005)
32. Yanover, C., Weiss, Y.: Finding the M most probable configurations using loopy belief propagation. In: Thrun, S., Saul, L., Schölkopf, B. (eds.) Advances in Neural Information Processing Systems 16, pp. 289–295. MIT Press, Cambridge (2004)

Chapter 7
Adaptive Evolutionary Algorithm Based on a Cliqued Gibbs Sampling over Graphical Markov Model Structure

Eunice Esther Ponce-de-Leon-Senti and Elva Diaz-Diaz

Abstract. This chapter introduces Estimation of Distribution Algorithms (EDAs) based on a learning strategy with two steps. The first step is based on the estimation of the searching sample complexity through an index based on the sample entropy. The searching sample algorithm learns a tree, and then, uses a sample complexity index to prognose the missing edges to obtain the cliques of the structure of the estimating distribution adding more edges if necessary. In the second step a new population is generated by a new cliqued Gibbs sampler (CG-Sampler) that drags through the space of solutions driven by the cliques of the learned graphical Markov model. Two variants of this algorithm are compared, the Adaptive Tree Cliqued - EDA (ATC-EDA) and the Adaptive Extended Tree Cliqued - EDA (AETC-EDA), and the Boltzmann selection procedure is used in CG-Sampler. They are tested with 5 known functions defined for 48, 50, 99 and 100 variables, and compared to Univariate Marginal Distribution Algorithm (UMDA). The performance of the two algorithms compared to UMDA is equal for OneMax and ZeroMax functions. The ATC-EDA and AETC-EDA are better than the UMDA for the other 3 functions.

7.1 Introduction

In this chapter, an estimation of distribution algorithm (EDA) is introduced based on undirected unrestricted graphical models (Markov random models). Two variants of EDAs are presented. In the first step the algorithms learn the structure of the solutions space represented by its cliques. This task is attained with two different learning options algorithms, the CL algorithm [3], and the extended tree algorithm (ETreeAL) presented in [4].

Eunice Esther Ponce-de-Leon-Senti · Elva Diaz-Diaz
Computer Sciences Department, Autonomous University of Aguascalientes,
Avenida Universidad 940 Ciudad Universitaria C.P. 20131 Aguascalientes, Mexico
e-mail: eponce@correo.uaa.mx, ediazd@correo.uaa.mx

S. Shakya and R. Santana (Eds.): Markov Networks in Evolutionary Computation, ALO 14, pp. 109–123.
springerlink.com

In the second step the algorithms generate samples using a new Gibbs sampler presented in this chapter. This sampling generator is named cliqued Gibbs sampler because in this sampler the variables are blocked by the cliques. In this case the Gibbs sampler generates samples from the Gibbsian distribution corresponding to the potentials given by the cliques obtained in the first step. This Gibbs sampler named cliqued Gibbs sampler (CG-sampler) obtains samples without knowing the parameters of the distribution (see Theorem 1) in Section 7.2.1. The cliqued Gibbs sampler uses a Boltzmann selection procedure [8]. This selection procedure is used to generate the samples from the models structures.

These algorithms have to solve two difficult tasks, one is the learning part that consists of determining the structure of the probability model and to estimate the parameters. The second difficult task is the sampling part. The sampling part constructs the population of solutions that at each step constitute a trail leaved by the movement of the algorithm in the space of solutions. The learning part is hard because the probabilistic model structure is unrestricted, and in this case the learning problem is exponentially complex in the number of variables. The parameters estimation is complex for unrestricted models, but in the proposed EDAs the estimation is realized only for the tree structure and only the prognosed structure is needed to give the input to the CG-sampler introduced in this chapter. The sampling part is difficult because the problem of obtaining a good sampling generator, capable of reflecting the dynamic of the search in the space of solutions, has not been conveniently solved until now. The results of these two new algorithms introduced are compared to the results of the UMDA by Mühlenbein and Paaß [11].

This chapter is divided into seven sections. Section 7.2 gives the theoretical foundations necessary to construct the algorithms and to assess the complexity of the problems. Section 7.3 gives the algorithms descriptions, and pseudocodes, such as, the CL algorithm, the extended tree adaptive learning algorithm (ETreeAL), the cliqued Gibbs sampler optimizer (CG- Sampler) and finally the adaptive tree cliqued estimation of distribution algorithm (ATC–EDA), and adaptive extended tree cliqued estimation of distribution algorithm (AETC–EDA). Section 7.4 describes and defines the test functions, section 7.5 describes the experiments, section 7.6 contains tables, results and discussion about the behavior of the two proposed algorithms, and a comparison with UMDA. Finally, the section 7.7 contains the conclusions and proposals for future work.

7.2 Concepts and Definitions

Because of the Gibbs distribution, and Markov random field equivalence (Theorem 1), it is possible to construct a Gibbs sampler optimizer based on knowing the cliques of the graphical model learned from the solutions space. This structure is learned evaluating the optimization function. The fundamental concepts, definitions and results to fulfill this task are given in this section.

7.2.1 Gibbs Distribution and Markov Random Fields (MRF)

Let $S = \{s_1, s_2, ..., s_v\}$ be a set of sites and let $\mathcal{G} = \{\mathcal{G}_s, s \in S\}$ be the neighborhood system for S, meaning it any collection of subset of S for which

1) $s \notin \mathcal{G}_s$ and
2) $s \in \mathcal{G}_r \Leftrightarrow r \in \mathcal{G}_s$.

$\mathcal{G}_s$ is the set of neighbors of s, and the pair $\{S, \mathcal{G}\}$ is a graph.

Definition 1. A subset $C \subset S$ is **a clique** if every pair of distinct sites in C are neighbors. $\mathcal{C} = \{C\}$ denotes the set of cliques. Let $X = \{x_s, s \in S\}$ denotes any family of random binary variables indexed by S. Let Ω be the set of all possible values of X, that is, $\Omega = \{w = (x_1, x_2, ..., x_v) : x_i \in \{0, 1\}\}$ is the sample space of all possible realizations of X.

Definition 2. X is **a Markov random field (MRF) with respect to** $\mathcal{G}$ if

$$P(X = w) > 0 \text{ for all } w \in \Omega \tag{7.1}$$

and

$$P(X_s = x_s | X_r = x_r, r \neq s) = P(X_s = x_s | X_r = x_r, r \in \mathcal{G}_s) \text{ for every } s \in S \text{ and } w \in \Omega \tag{7.2}$$

where X denotes the random variable and w denotes the values that this variable can take.

Definition 3. A Gibbs distribution relative to $\{S, \mathcal{G}\}$ is a probability measure π on Ω with the following representation

$$\pi(w) = \frac{1}{Z} e^{-U(w)/T} \tag{7.3}$$

where Z and T are constants. U is called the energy function and has the form

$$U(w) = \sum_{c \in \mathcal{C}} V_c(w) \tag{7.4}$$

Each V_c is a function on Ω that only depends on the coordinates x_s of w for which $s \in C$, and $C \subset S$.

Definition 4. The family $\{V_c, c \in \mathcal{C}\}$ is called **a potential** and the constant

$$Z = \sum_{w} e^{-U(w)/T} \tag{7.5}$$

is called **the partition function**. The constant T is named temperature and it controls the degree of "peaking" in the density π.

Theorem 1. *(Equivalence theorem) Let $\mathcal{G}$ be a neighborhood system. Then X is a MRF with respect to $\mathcal{G}$ if and only if $\pi(w) = P(X = w)$ is a Gibbs distribution with respect to $\mathcal{G}$.*

A more extensive treatment can be seen in [5], [2], [6].

7.2.2 Gibbs Sampling

Gibbs Sampling is a Markovian updating scheme that works as follows. Given an arbitrary starting set of values $x^{(0)} = (x_1^{(0)}, x_2^{(0)}, ..., x_v^{(0)}) \in \Omega$, one value $x_i^{(0)}$ is drawn from the conditional distribution $P(x_i | x_1^{(0)}, ..., x_{i-1}^{(0)}, x_{i+1}^{(0)}, ..., x_v^{(0)})$ for each $i = 1, ..., v$, where v is the number of variables. So, each variable is visited in the natural order until v. After that a new individual $x^{(1)} = (x_1^{(1)}, x_2^{(1)}, ..., x_v^{(1)})$ is obtained.

Geman and Geman [5] demostrated that for

$$t \to \infty, \ \ x^{(t)} \to x$$

where $x = (x_1, x_2, ..., x_v)$ and t is the parameter of the process (if the process is an algorithm, t is an iteration). This sampling schema required v random variate generations, one for each state i of the schema.

7.2.3 Selection Procedure

As a component of the cliqued Gibbs sampler, the generator part, a selection procedure is used.

Definition 5. Let x_i be a solution and $f(x_i)$ be the function of x_i to optimize, then the **Bolztmann selection** procedure evaluates a solution x_i

$$P(x_i) = \frac{\exp(-f(x_i))}{E(x_m)} \tag{7.6}$$

where $E(x_m) = \sum_j f(x_j) p(x_j)$ is the expected value of all the elements in the current population.

$P(x_i)$ is not invariant to scaling but is invariant to translation [8].

7.2.4 Random Selection Schema

Given the state i of the schema $x^{(t)}$, for the value $P[x_i | x_1^{(t)}, ..., x_{i-1}^{(t)}, x_{i+1}^{(t)}, ..., x_v^{(t)}]$ a value for $x_i^{(t+1)}$ is obtained selecting as follows: if $p(x_i^{(t+1)}) > p(x_i^{(t)})$ select $x_i^{(t+1)}$, else calculate

$$q = \frac{p(x_i^{(t+1)})}{p(x_i^{(t)})}, \tag{7.7}$$

and if a random number uniformly generated in the interval $[0, 1]$ is less that q, select $x_i^{(t+1)}$.

7.2.5 The Kullback-Leibler Divergence

Definition 6. The **K-L divergence from the probability model M to the data x** is given by the Kullback-Leibler information [1]

$$G^2(M,x) = \log(L(\widehat{m}_n^M(x)) = -2\sum_{i=1}^{k} x_i \log_2(\frac{\widehat{m}_i^M}{x_i}) . \tag{7.8}$$

where k is the sample number of different individuals, n is the total number of individuals, and $\widehat{m}_n^M$ is the maximum-likelihood parameter estimator of m_n^M.

The K-L divergence is also known as relative entropy, and can be interpreted as the amount of information in the sample x not explained by the model M, or the deviation from the model M to the data x. This K-L divergence is used to calculate $SMCI$ (Definition 8) and in time $EMUBI$ (Definition 9) in the next paragraphs.

Definition 7. The **mutual information measure** $I_{X_iX_j}$ for all $X_i, X_j \in X$ is defined as

$$I_{X_iX_j} = I(X_i,X_j) = \sum_{x_i,x_j} P(x_i,x_j) \log \frac{P(x_i,x_j)}{P(x_i)P(x_j)} . \tag{7.9}$$

where $P(x_i,x_j) = P(X_i = x_i, X_j = x_j)$.

7.2.6 Statistical Model Complexity Index

As a part of a strategy to learn a graphical Markov model, a statistical model complexity index ($SMCI$) is defined and tested by Diaz et al. [4]. Based on this index it is possible to obtain an evaluation of the sample complexity and to prognose the graphical model to explain the information contained in the sample.

The model representing the uniform distribution is denoted by M_0, and the model represented by a tree is denoted by M_T. If a sample is generated by a model M containing more edges than a tree, the information about the model M contained in this sample and not explained, when the model structure is approximated by a tree, may be assessed by the index defined as follows.

Definition 8. Let x be a sample generated by a model M. The **statistical model complexity index ($SMCI$) of the model M** [4] is defined by the quantitative expression

$$SMCI(M,x \mid M_T) = \frac{G^2(M_T,x) - G^2(M,x)}{G^2(M_0,x)} . \tag{7.10}$$

This index can be named sample complexity index, because it is assessed by the quantity of information contained in the sample generated by the model M.

Definition 9. Let G be the graph of the model M, and let v be the number of vertices, let $MNE(v)$ be the maximum number of edges formed with v vertices. The **edge missing upper bound index ($EMUBI$)** (see [4]) is defined by

$$EMUBI_\tau(M,x \mid M_T) = \tau(MNE(v) - v + 1)SMCI(M,x \mid M_T) \,. \tag{7.11}$$

where τ is the window allowing a proportion of variability in the sample to get into the model [4]. This coefficient is a filter that allows to sample relevant information for the model structure construction.

This index is used to prognose the number of edges to add to a tree in order to approximate the complexity of the sample using the graphical model, in this case it is an unrestricted graphical Markov model.

7.3 Algorithms Descriptions and Pseudocodes

In this section a description of each of the algorithms is given. Some of them are already presented in other papers: the CL algorithm in [3], Extended tree algorithm (ETreeAL) in [4]. The Cliqued Gibbs Sampler (CG – Sampler) Optimizer and the two types of Adaptive Tree and Extended Tree Cliqued – EDA (ATC- EDA with Bolztmann selection, and AETC-EDA with Bolztmann selection) are new types of discrete EDAs, presented for the first time in this chapter.

7.3.1 *Chow and Liu Maximum Weight Spanning Tree Algorithm (CL)*

The CL algorithm (Chow and Liu algorithm) [3] obtains the maximum weight spanning tree using the Kruskal algorithm [7] and the mutual information values $I_{X_iX_j}$ (Definition 7) for the random variables. The tree obtained by this algorithm is denoted by $M_T(CL)$.

Algorithm 1. CL algorithm

Input: Distribution P over the random vector $X = (X_1, X_2, ..., X_v)$

1 Compute marginal distributions P_{X_i} , $P_{X_iX_j}$, for all $X_i, X_j \in X$.

2 Compute mutual information values (Definition 7) $I_{X_iX_j}$ for all $X_i, X_j \in X$.

3 Order the values from high to low (w.r.t.) mutual information.

4 Obtain the maximum weight spanning tree $M_T(CL)$ by the Kruskal algorithm [7].

Output: The maximum weight spanning tree $M_T(CL)$

7.3.2 *Extended Tree Algorithm (ETreeAL)*

In this section the tree $M_T(CL)$ obtained by CL algorithm (Section 7.3.1) is extended. A number of edges is added to $M_T(CL)$ based on EMUBI (Definition 9). The number of edges added, try to adapt the complexity of the learning model to the data complexity, as described in the following algorithm. This algorithm has a parameter τ that plays a regularization role to avoid overfitting the model to the sample. This algorithm was presented and tested in [4]. It finds a tree as Chow and

Liu and extends it using the $EMUBI$. The output $M_{EXT}(CL)$ of ETreeAL algorithm is the structure of the model given by its cliques.

The CL-algorithm calculates for the number of variables v, $\frac{v(v-1)}{2}$ mutual informations. If each variable takes two values then the CL-algorithm has complexity $O(\frac{v(v-1)}{2}2^2) = O(2v(v-1)) = O(v^2)$. The ETreeAL adds the number of prognosed edges to the tree to obtain the model structure necessary to the CG-sampler input, and these operations have polynomial complexity.

Algorithm 2. The extended tree adaptive learning algorithm (ETreeAL)

Input: Distribution P over the random vector $X = (X_1, X_2, ..., X_v)$

1 Call CL Algorithm in order to obtain the maximum spanning tree using the mutual information measure $I_{X_iX_j}$ (Definition 7) as weight edges.
2 Calculate the edge missing upper bound prediction index $(EMUBI_\tau(M, x \mid M_T(CL)))$.
3 Add to $M_T(CL)$, τ percent from missing edges in the order of the mutual information values (Definition 7).

Output: Extended tree model structure $M_{EXT}(CL)$

7.3.3 Cliqued Gibbs Sampler (CG–Sampler) Optimizer

This algorithm needs the cliques of the graphical models as input to fulfill the condition of equivalence required by theorem 1 (see Section 7.2.1). The variables of each clique are used together by the CG– Gibbs sampler optimizer to generate new individuals for the sample. Other authors used blocked Gibbs sampler [13], and that is why the name "Cliqued Gibbs Sampler" was given to this sampler. Let $f(x)$ be the objective function. $\widehat{E}(x_m)$ is the $f(x)$ mean estimator. Without losing generality in the description of the CG-Sampler algorithm, we assume that the optimization problem is to obtain a minimum and a convenient change can be made to obtain a maximum.

Observing Algorithm 3 it is seen that the algorithm input consists of a population of selected solutions, let P be the population and let the structure of the graphical model adjusted be given by its cliques, $\mathscr{C} = \{c_1, ..., c_k\}$. In the outer cycle the iterations run through all elements of the population (N). The first inner **for** runs through the number of cliques (k) and the step 7 generates a marginal table for each clique and calculates a roulette selection running over the marginal table corresponding to the clique c_j for all j=1,..,k. Let $CM = \max_{c_j \in \mathscr{C}} |c_j|$ then the worst case is when $|c_j| = CM$ for all j. In this case the complexity is given by $O((N)(k)2^{CM})$, so the complexity of the CG-Sampler is exponential in the size of the maximum clique of the model.

7.3.4 Adaptive Extended Tree Cliqued – EDA (AETC–EDA)

The AETC - EDA employs the ETreeAl to obtain the Markov model structure of the population of solutions for each algorithm's iteration, with this structure the CG- sampler optimizer obtains the next population.

Algorithm 3. CG–Sampler Optimizer Algorithm

```
Input: A population of selected solutions, and the structure of the graphical model,
       given by its cliques
1  P_New ← {}
2  i ← 1
3  repeat
4      Take the solution i of the selected solutions
5      for j = 1 to Number of cliques do
6          Take the clique j
7          Generate the marginal table of the clique j
8          Roulette Selection: Select the individual corresponding to the k cell of the
           marginal table according to its selection probability (Boltzmann Selection)
           Definition 5.
9      end
10     if Ê(x_m) < f(k) then
11         P_New ← P_New ∪ {k}
12     end
13     else
14         /* Metropolis Step */
15         if Random ≤ f(k)/Ê(x_m) then
16             P_New ← P_New ∪ {k}
17         end
18         else
19             P_New ← P_New ∪ {i}
20         end
21     end
22     if i =(Percent of selection)*N/100 then
23         i = 0
24     end
25     else
26         i = i + 1
27     end
28 until (a new population P_New with size N is obtained) ;
Output: A new population P_New (Gibbsian Population)
```

7.3.5 *Adaptive Tree Cliqued – EDA (ATC–EDA)*

The ATC-EDA uses the tree structure obtained by the CL algorithm to guide the cliqued Gibbs sampler optimizer.

7.4 Test Functions

Five functions are employed to test the two algorithms presented in this chapter and the UMDA. The selected functions are a sample from functions of different

Algorithm 4. Adaptive Extended Tree Cliqued – EDA (AETC–EDA)

Input: Number of variables, function to optimize, population size, percent of selected individuals, stop criterion
1 Create the initial population of size N at random.
2 **repeat**
3 Evaluate the population.
4 Order the population and select a portion.
5 With the selected portion of the population call the *ETreeAl* algorithm (Section 7.3.2).
6 Call cliqued Gibbs sampler (CG–Sampler) optimizer (Section 7.3.3)
7 **until** *(Some stop criterion is met)* ;
Output: Solution of the optimization problem

Algorithm 5. Adaptive Tree Cliqued – EDA (ATC–EDA)

Input: Number of variables, function to optimize, population size, percent of selected individuals, stop criterion
1 Create the initial population of size N at random.
2 **repeat**
3 Evaluate the population
4 Order the population and select a portion
5 With the selected portion of the population calls the CL algorithm (Section 7.3.1)
6 Call cliqued Gibbs sampler (CG–Sampler) optimizer (Section 7.3.3)
7 **until** *(Some stop criterion is met)* ;
Output: Solution of the optimization problem

difficulties. Some of them are deceptive functions proposed to study the genetic algorithm performance.

In all functions we use v as the number of the variables, x_i is a binary variable for every i, and $u = \sum x_i$ is the number of ones in the solution $\mathbf{x} = (x_1, ..., x_v)$.

OneMax Problem. This is a well-known simple linear optimization problem used to test the performance and the convergence velocity. It is defined as maximizing the function

$$F_{OneMax}(\mathbf{x}) = \sum_{i=1}^{v} x_i \tag{7.12}$$

ZeroMax Problem. This is defined as maximizing the function where $u = \sum x_i$ is the number of zeros in the solution $\mathbf{x} = (x_1, ..., x_v)$.

$$F_{ZeroMax}(\mathbf{x}) = \sum_{i=1}^{v} x_i \tag{7.13}$$

$\mathbf{F_{c_2}}$ Deceptive Problem. Proposed in [10] its auxiliary function and deceptive decomposable function are as follows.

$$f_{Muhl}^{5} = \begin{cases} 3.0 \text{ for } x = (0,0,0,0,1) \\ 2.0 \text{ for } x = (0,0,0,1,1) \\ 1.0 \text{ for } x = (0,0,1,1,1) \\ 3.5 \text{ for } x = (1,1,1,1,1) \\ 4.0 \text{ for } x = (0,0,0,0,0) \\ 0.0 \text{ otherwise} \end{cases}$$

$$f_{c_2}(\mathbf{x}) = \sum_{i=1}^{\frac{v}{5}} f_{Muhl}^{5}(x_{5i-4}, x_{5i-3}, x_{5i-2}, x_{5i-1}, x_{5i}) \tag{7.14}$$

$\mathbf{F_3}$ Deceptive Problem. This problem has been proposed in [10]. Its auxiliary function and deceptive decomposable function are as follows.

$$f_{dec}^{3} = \begin{cases} 2 \text{ for } u = 0 \\ 1 \text{ for } u = 1 \\ 0 \text{ for } u = 2 \\ 3 \text{ for } u = 3 \end{cases}$$

$$f_{3deceptive}(\mathbf{x}) = \sum_{i=1}^{\frac{v}{3}} f_{dec}^{3}(x_{3i-2}, x_{3i-1}, x_{3i}) \tag{7.15}$$

$\mathbf{Trap_k}$ Problem. A Trap function of order k [12] can be defined as

$$Trap_k(u) = \begin{cases} k & u = k \\ k - 1 - u, & otherwise \end{cases}$$

$$f_{Trap_k}(\mathbf{x}) = \sum_{i=1}^{\frac{v}{k}} Trap_k(x_{5i-4}, x_{5i-3}, x_{5i-2}, x_{5i-1}, x_{5i}) \tag{7.16}$$

We use $k = 5$.

7.5 Experiments

In this section the experiment performed is described, the ATC – EDA, AETC – EDA, and the UMDA are tested and compared. The algorithms are tested with the functions: OneMax, ZeroMax, F_{c_2}, F_3 Deceptive and $Trap_5$ as defined in the previous paragraph. These functions are constructed with vectors of 48, 50, 99 and 100 variables. For 48 and 50 variables the population contains 100 and 500 individuals, and for 99 and 100 variables the population contains 800 and 1500 individuals. Percent of selection is fixed at 40 for all problems with 48 and 50 variables and the percent of selection is fixed at 60 for all problems with 99 and 100 variables. The stopping rule has two criteria: until the algorithm obtains the optimum or until the number of evaluations is $N * 1001$, where N is the population size. For each test function, each algorithm runs 30 replications.

7.6 Results and Discussion

The results are presented in Tables from 7.1 to 7.6 for the functions OneMax, ZeroMax, F_{c_2}, F_3 Deceptive and $Trap_5$. In Table 7.1, the results of algorithm ATC – EDA for 48 and 50 variables, without Metropolis step for the five instances of those functions are presented. In the case of AETC–EDA for 48 and 50 variables (Table 7.2), a $\tau = 0.67$ without Metropolis step is used. In Table 7.3, the results of algorithm AETC – EDA for 99 and 100 variables, with Metropolis step and $\tau = 0.67$, for the same five instances of those functions are presented. In Tables from 7.1 to 7.3, the CG- Sampler with Bolztmann selection is used. In Tables 7.4 for 48 and 50 variables, and Tables 7.5 and 7.6 for 99 and 100 variables, the results of the UMDA [9] using the same five instances of the problems are presented. For each type of function the number of variables, the population size, the average number of function evaluations, the average number of cliques in models, the average clique sizes, the mean best value and the average time are reported. The last five numbers are obtained averaging the results of the 30 replications.

The OneMax, ZeroMax, F_{c_2}, F_3 Deceptive and $Trap_5$ for the number of variables ≤ 50 can be solved using the ATC-EDA (the tree model). In this case the CG Sampler provides the variability to help the algorithm to escape from the local optimum, and does not need the Metropolis step (see Table 7.1). The AETC–EDA for ≤ 50 variables and population size 100 was tested to know if this algorithm can obtain the optimum, and it is seen that the answer is yes, it obtains the optimum in all problems. In this case the algorithm used a lot of cliques with more than two vertices (see Table 7.2).

The same list of problems in the case of 99, 100 variables needs the Metropolis step, to get out of local optimum (see Table 7.3). For the number of variables greater than 50 and less or equal than 100 the ATC-EDA runs very slowly and it was necessary to use the AETC-EDA that permits graphical models more complex than a tree, that is, cliques with more than 2 vertices. In the case of F_3 Deceptive problem, some cliques with 3 vertices represents the topology of the F_3 Deceptive function. For the case of 99, and 100 variables it is necessary to close the window of information using $\tau = 0.67$ (see Table 7.3).

The UMDA for 50 variables obtains the optimum in the problems OneMax and ZeroMax for all replications. For F_{c_2} the UMDA obtains the optimum for some proportion of the replications. For the F_3 Deceptive and $Trap_5$, the mean best values are short to the optimum value (see Table 7.4). The UMDA for 99 and 100 variables obtains the optimum in the problems OneMax and ZeroMax for all replications. For F_3 Deceptive problem the UMDA obtains the optimum for some high per cent of replications. For the F_{c_2} and $Trap_5$, the mean best values are short to the optimum value (see Tables 7.5 and 7.6).

The performance of the two algorithms presented and the UMDA for 48 and 50 differ greatly for F_{c_2}, F_3 Deceptive and $Trap_5$ in the number of evaluations, and in the execution time. Besides the UMDA obtains the optimum only 16 times for F_{c_2}, and 2 times for F_3 Deceptive, but never for the $Trap_5$. For 99 and 100 variables

Table 7.1 ATC – EDA (CG - Sampler with Bolztmann Selection without Metropolis Step). Percent of Selection= 40. Number of Generations= 1000. The algorithm obtains the optimum in all cases. All cliques have two vertices (variables).

Parameters	OneMax	ZeroMax	F_{c_2}	F_3 Deceptive	$Trap_5$
Variables	50	50	50	48	50
Population size	500	500	500	500	500
Evaluations Number	6060.5333	6729.7333	3283.7667	3386.6667	19832.2
Cliques Numbers	49	49	49	47	49
The Optimum Value	50	50	40	48	50
Mean Best Value	50	50	40	48	50
Seconds	1.2464	1.3804	1.4987	1.2083	1.0254

Table 7.2 AETC – EDA (CG - Sampler with Bolztmann Selection and without Metropolis Step). Percent of Selection= 40. $\tau = 0.67$. Number of Generations= 1000. The algorithm obtains the optimum in all cases.

Parameters	OneMax	ZeroMax	F_{c_2}	F_3 Deceptive	$Trap_5$
Variables	50	50	50	48	50
Population size	100	100	100	100	100
Evaluations Number	5218.4667	640.4333	771.1667	8583	35581.4333
Cliques Numbers	88.7333	76.6333	48.0333	64.8667	67.2
Cliques Size	2.6271	2.4622	2.4357	2.6918	2.8559
The Optimum value	50	50	40	48	50
Mean best value	50	50	40	48	50
Seconds	3.9554	0.4917	0.655	9.9066	126.0498

Table 7.3 AETC – EDA (CG - Sampler with Bolztmann Selection and Metropolis Step). Percent of Selection= 60. $\tau = 0.67$. Number of Generations= 1000. The algorithm obtains the optimum in all cases.

Parameters	OneMax	ZeroMax	F_{c_2}	F_3 Deceptive	$Trap_5$
Variables	100	100	100	99	100
Population size	800	800	800	800	800
Evaluations Number	29270	30102.5333	303074	76534.6	19832.2
Cliques Numbers	103.7667	102.0667	98	95.1333	97.1333
Cliques Size	2.0081	2.0119	2.0102	2.0315	2.022
The Optimum value	100	100	80	99	100
Mean best value	100	100	80	99	100
Seconds	30.2073	51.6994	1791.5799	133.4775	31.5074

Table 7.4 UMDA. Percent of Selection= 40. Number of Generations= 1000.

Parameters	OneMax	ZeroMax	F_{c_2}	F_3 Deceptive	$Trap_5$
Variables	50	50	50	48	50
Population size	500	500	500	500	500
Evaluations Number	3239.8333	3608.0333	238491.2333	467576.7	500500
The Optimum value	50	50	40	48	50
Mean best value	50	50	39.5333	45.8667	36
Optimum obtained (number)	30	30	16	2	0
Seconds	0.2506	0.2886	20.3493	54.4991	48.2368

Table 7.5 UMDA. Percent of Selection= 60. Number of Generations= 1000.

Parameters	OneMax	ZeroMax	F_3 Deceptive	F_{c_2}	$Trap_5$
Variables	100	100	99	100	100
Population size	800	800	800	800	800
Evaluations Number	11075.4667	16269.2667	334816.7333	800800	800800
The Optimum value	100	100	99	80	100
Mean best value	100	100	98.4667	70.1	76
Optimum obtained (number)	30	30	18	0	0
Seconds	1.0762	3.0934	13.5078	114.459	114.8956

Table 7.6 UMDA. Percent of Selection= 60. Number of Generations= 1000.

Parameters	OneMax	ZeroMax	F_3 Deceptive	F_{c_2}	$Trap_5$
Variables	100	100	99	100	100
Population size	1500	1500	1500	1500	1500
Evaluations Number	20505.8333	29712.7	142297.1667	1501500	1501500
The Optimum value	100	100	99	80	100
Mean best value	100	100	98.9333	71.25	76
Optimum obtained (number)	30	30	28	0	0
Seconds	3.9281	3.7273	14.7484	115.138	143.1661

the UMDA never finds the optimum for F_{c_2} and $Trap_5$ neither for 800 nor for 1500 population size.

7.7 Conclusions and Proposals for Future Work

The use of the equivalence theorem (Theorem 1) supports the use of the cliqued Gibbs Sampler, and the two EDAs algorithms obtain the optimum in all problems employed as test, adapting the performance to the complexity of the objective functions. Compared with the UMDA the running time is always shorter and the new EDAs always attain the optimum solutions for each of the five objective functions, while the UMDA only attains this performance for the One Max and the Zero Max problems. Observing the number of cliques, the objective functions F_3 Deceptive, F_{c_2} and $Trap_5$ are represented by models two or four cliques with more than two vertices containing. In particular for the F_3 Deceptive function the algorithm represented by cliques with 3 vertices fast all (except 3) groups of three variables present in the objective function.

The future work will be directed to the analysis of the complexity if the sample solutions space measured by the model sample complexity, and the classification of the objective function problems, relating it to the graphical model learned to support the Cliqued Gibbs Sampler. As future work these algorithms will be used with more complex functions, and it will be analyzed the relation between the functions complexity and the difficulty for the EDAs to find the optimal solution.

Acknowledgements. We would like to acknowledge support for this project (PIINF10-2) from the Autonomous University of Aguascalientes, Aguascalientes, Mexico. The reviewers of the chapter are greatly acknowledged for their constructive comments.

References

1. Akaike, H.: A new look at the statistical model identification. IEEE Transactions on Automatic Control 19(6), 716–723 (1974)
2. Besag, J.E.: Spatial interaction and the statistical analysis of lattice systems (with discussion). J. Royal Statist. Soc. Series B 36, 192–326 (1974)
3. Chow, C.K., Liu, C.N.: Approximating discrete probability distributions with dependence trees. IEEE Transactions on Information Theory IT-14(3), 462–467 (1968)
4. Diaz, E., Ponce-de-Leon, E., Larrañaga, P., Bielza, C.: Probabilistic Graphical Markov Model Learning: An Adaptive Strategy. In: Aguirre, A.H., Borja, R.M., Garciá, C.A.R. (eds.) MICAI 2009. LNCS, vol. 5845, pp. 225–236. Springer, Heidelberg (2009)
5. Geman, S., Geman, D.: Stochastic relaxation, Gibbs distributions and the bayesian distribution of images. IEEE Transactions on Pattern Analysis and Machine Intelligence 6, 721–741 (1984)
6. Kindermann, R., Snell, J.L.: Markov random fields and their applications. Contemporary Mathematics. American Mathematical Society, Providence, RI (1980)
7. Kruskal, J.B.: On the shortest spanning tree of a graph and the traveling salesman problem. In: Proceeding American Mathematical Society, vol. 7, pp. 48–50. American Mathematical Society (1956)
8. de la Maza, M., Tidor, B.: An analysis of selection procedures with particular attention paid to proportional and Boltzmann selection. In: Proceedings of the 5th International Conference on Genetic Algorithms, pp. 124–131. Morgan Kaufmann Publishers Inc., San Francisco (1993)
9. Mühlenbein, H.: The equation for the response to selection and its use for prediction. Evolutionary Computation 5(3), 303–346 (1998)

10. Mühlenbein, H., Mahnig, T., Ochoa Rodriguez, A.: Schemata, distributions and graphical models in evolutionary optimization. Journal of Heuristic 5, 215–247 (1999)
11. Mühlenbein, H., Paaß, G.: From Recombination of Genes to the Estimation of Distributions I. Binary Parameters. In: Ebeling, W., Rechenberg, I., Voigt, H.-M., Schwefel, H.-P. (eds.) PPSN 1996. LNCS, vol. 1141, pp. 178–187. Springer, Heidelberg (1996)
12. Pelikan, M.: Bayesian Optimization Algorithm: From Single Level to Hierarchy. PhD thesis, University Illinois at Urbana Champain, Also IlliGAL Report No. 2002023 (2002)
13. Santana, R., Mühlenbein, H.: Blocked stochastic sampling versus estimation of distribution algorithms. In: Proceedings of the 2002 Congress on the Evolutionary Computation, CEC 2002, vol. 2, pp. 1390–1395. IEEE Press (2002)

Chapter 8
The Markov Network Fitness Model

Alexander E.I. Brownlee, John A.W. McCall, and Siddhartha K. Shakya

Abstract. Fitness modelling is an area of research which has recently received much interest among the evolutionary computing community. Fitness models can improve the efficiency of optimisation through direct sampling to generate new solutions, guiding of traditional genetic operators or as surrogates for a noisy or long-running fitness functions. In this chapter we discuss the application of Markov networks to fitness modelling of black-box functions within evolutionary computation, accompanied by discussion on the relationship between Markov networks and Walsh analysis of fitness functions. We review alternative fitness modelling and approximation techniques and draw comparisons with the Markov network approach. We discuss the applicability of Markov networks as fitness surrogates which may be used for constructing guided operators or more general hybrid algorithms. We conclude with some observations and issues which arise from work conducted in this area so far.

8.1 Introduction, Motivation and Alternatives

The end-goal of an evolutionary algorithm is usually efficient optimisation. Fitness modelling [20] is one of the many techniques which can support this goal, by reducing overall run-time or by simplifying the problem in some way.

Alexander E.I. Brownlee
Loughborough University, Loughborough, UK
e-mail: a.e.i.brownlee@lboro.ac.uk

John A.W. McCall
Robert Gordon University, Aberdeen, UK
e-mail: j.mccall@rgu.ac.uk

Siddhartha K. Shakya
Business Modelling and Operational Transformation Practice,
BT Innovate & Design, Ipswich, UK
e-mail: sid.shakya@bt.com

S. Shakya and R. Santana (Eds.): Markov Networks in Evolutionary Computation, ALO 14, pp. 125–140.
springerlink.com

Previous work including [4, 50–52] has described a number of approaches using undirected probabilistic graphical models (Markov networks) within a framework called Distribution Estimation Using Markov networks (DEUM), an Estimation of Distribution algorithm [24]. With the Markov network approach model-building overhead is significant, particularly as the number of variables and interactions increases. While the number of function evaluations to reach an optimum were significantly fewer than with other algorithms in our earlier work the time taken to build and sample the probabilistic model was large. However, it has been found that the Markov network at the heart of DEUM represents a good model of fitness [5, 7], and this has been exploited in recent work [8] as well as providing an explanation for the previous good results.

The broader concepts of fitness modelling, approximation and surrogates [20, 26] have attracted much interest among the evolutionary computing community. A motivation for using fitness modelling within an evolutionary algorithm is to improve the efficiency of optimisation; this can be achieved in several ways. If evaluating the model is cheaper than evaluating the fitness function, it can be used to reduce overall run-time [21, 31, 41, 57, 62]. A model may be used where no explicit fitness function exists such as in evolutionary art and music [22]. Further, a fitness model may be employed to simplify the search by reducing noise [10, 42, 58] or smoothing a multimodal landscape [63].

Many approaches to fitness modelling exist. One common approach is artificial neural networks [21], with more recent examples including [57] and [14]. In [33] an algorithm groups individuals of similar fitness into classes that are then passed to Bayesian classifiers which can be sampled to generate individuals of high fitness. Schmidt and Lipson [48] use co-evolution to generate fitness predictors. In [12], an archive of already-evaluated solutions are fuzzily matched to new solutions, with fitness taken from the archive if matches are found. The Learnable Evolution Model (LEM) [32] incorporates machine learning to identify features distinguishing high and low fitness individuals. In [13], the authors report the use of a Gaussian random field meta-model as a surrogate in a (μ+λ) ES for single and multi-objective continuous problems with good results on a number of benchmarks.

Fitness inheritance (passing of fitness values from parents to offspring) to reduce the number of fitness evaluations [11, 56] is arguably a form of fitness modelling. A fitness model may also be used to guide standard genetic operators such as crossover and mutation as in [1, 21, 27, 40, 60]. Other hybrid approaches combine probabilistic models with different algorithms such as that described by [36, 47, 61]. In contrast to the use of undirected models in this chapter there are has also been some work done using directed probabilistic graphical models for fitness modelling, such as that described as a variant of fitness inheritance in [38].

Polynomial regression or the fitting of a response surface has also been used to construct a model of fitness [64]. The Markov fitness model described in this chapter bears some similarity to this, in effect being a response surface for the fitness function.

Much of the above work concentrates on continuous fitness functions. The Markov fitness model described here models discrete functions and there are a number of related works with such functions. One of the earliest is [35], where the authors use a neural network to classify low and high fitness solutions with a bit-string encoding. In [65], the authors use a meta-model built using machine learning techniques (one example is genetic programming), also to classify high and low fitness solutions having a bit-string encoding. Some approaches map the discrete function onto a continuous space: in [25], a radial basis function network is used as a surrogate for a mixed-integer search space. The model is used as a filter; too many offspring are created each generation and the model is used to choose promising ones which are evaluated with the true fitness function and retained. Similar to this, [59] proposes candidate over sampling in EBCOA [33]; generating too many solutions and trimming back to the number required. The authors found that picking the solutions predicted by the model to be less fit worked best. Their work looked at bit-string encoded functions and used machine learning approaches to infer rules distinguishing high and low fitness solutions.

Han and Yang [18] describe mapping discrete variables onto continuous ones to allow multiple linear regression for screening variables prior to optimisation. In [30], Gaussian processes are used for learning discrete fitness landscapes, demonstrated on Multidimensional Gaussian Landscapes and NK landscapes. Takahashi et al [58] use fitness estimation from a statistical model of the history of solutions to deal with noisy fitness functions; the example given is on a weight vector with discrete values. This builds on the earlier work with continuous functions in [42]. Finally, in [34] the author builds a surrogate from a Gibbs model which is derived from the distribution learnt by an EDA. This is demonstrated with both discrete and continuous benchmark functions.

Jin's comprehensive 2005 review [20] presents a wider survey of existing work in this area, and further recent developments can be found in [55].

8.2 Defining the Model

Previous publications on DEUM including [5, 54] describe how a Markov network is used to model the distribution of energy across the set of variables in a bit-string encoded problem. In this section we summarise how the model is derived.

A Markov Network is a pair (G, Ψ), where G is the structure and the Ψ is the parameter set of the network. G is an undirected graph where each node corresponds to a random variable in the modelled data set and each edge corresponds to a probabilistic joint dependency between variables. We say that two nodes connected by an edge are *neighbouring nodes*. A subset $K = \{X_{i1}, ..., X_{ik}\}$ of k mutually-neighbouring nodes is termed a *k-clique*. Note that we include the empty set $\varnothing$ as a 0-clique and each singleton 1-clique $\{x_i\}$ in our definition of clique.

The Hammersley-Clifford Theorem states that the joint probability distribution of a Markov Network factorises as a Gibbs Distribution, completely determined by a set Ψ of parameters, each of which is a real number α_k that defines the energy contribution from clique k.

The precise form of the Gibbs Distribution is given in equation (8.1):

$$p(x) = \frac{e^{-U(x)/T}}{Z} \tag{8.1}$$

where,

$$Z = \sum_{y \in \Omega} e^{-U(y)/T} \tag{8.2}$$

Here, the numerator, $e^{-U(x)/T}$, represents the likelihood of a particular configuration x of the variables. The denominator, Z, is the normalising constant computed by summing over the set Ω of all possible configurations (note Z is never computed in practice). $U(x)$ is an *energy function* computed by summing contributions calculated from the values that x takes on each clique. Thus this exponentiated sum gives a factorisation of the distribution based on the structure G. We will consider the energy function in more detail shortly. T is a temperature constant which controls the ruggedness of the probability distribution.

The key idea of the DEUM EDA is to model solution fitness as a mass distribution that equates to the Gibbs distribution as shown in equation (8.3):

$$p(x) = \frac{f(x)}{\Sigma_{y \in \Omega} f(y)} = \frac{e^{-U(x)/T}}{Z} \tag{8.3}$$

Sampling this distribution will generate high fitness individuals with high probability. We now explain how the DEUM algorithm estimates this distribution.

Identifying corresponding terms in the numerator and denominator gives, for each solution $x = \{x_1, x_2, ..., x_n\}$, the following negative log relationship between $f(x)$ and $U(x)$:

$$-ln(f(x)) = U(x) \tag{8.4}$$

Let structure G contain a set of cliques $K = \{K_1, ..., K_m\}$. Then, for any solution $x = \{x_1, x_2, ..., x_n\}$, $U(x)$ has the form:

$$U(x) = \sum_i \alpha_i V_{K_i}(x) \tag{8.5}$$

The V_k are the characteristic functions of a Walsh decomposition of the fitness function. Walsh functions [3, 15–17] are a set of rectangular waveforms taking two amplitude values, +1 and -1. Similar to the use of Fourier transforms representing for analogue waveforms, Walsh functions may be combined linearly to represent any fitness function based on a bit-string representation.

The $V_K(x)$ for each clique K are defined in (8.6) to (8.8).

$$K = \varnothing \qquad V_{\varnothing}(x) \equiv 1 \ \forall \, x \tag{8.6}$$

$$K = \{i\} \qquad V_i(x) = \begin{cases} 1 & x_i = 1 \\ -1 & x_i = 0 \end{cases} \tag{8.7}$$

$$\text{For } K \subseteq |1,\ldots,n|, \quad |K| \geqslant 2, \quad V_K(x) = \prod_{i \in K} V_i(x) \tag{8.8}$$

Thus the energy function, and hence the fitness, is completely determined by the parameters α_i. The α_i are non-zero only for cliques present in the structure G. Given a sufficiently-sized population of solutions and their fitnesses, equations (8.4) and (8.5) yield a system of equations in the parameters that can be solved using a least-squares approximation to estimate the distribution. [50] describes how singular value decomposition [39] is used for this.

Of principle interest here is that, once the parameters are determined, we can combine (8.4) and (8.5) to obtain a model of the fitness function:

$$-ln(f(x)) = \sum_i \alpha_i V_{K_i}(x) \tag{8.9}$$

We call this the *Markov Fitness Model* (MFM) of f. We now proceed to discuss how the quality of the model may be measured before moving on to applications.

8.2.1 Model Quality and the Fitness Prediction Correlation

Given the relationship in (8.9), it is possible to extrapolate information about the fitness function and in particular the optimal solution by looking at the values given to the model parameters α_i. Minimising the energy of a solution is equivalent to maximising fitness. For the univariate terms this means that a positive α_i will require a negative value for $V(x_i)$ to minimise the energy contribution from that term. This equates to x_i (the ith bit) being set to 0. Likewise, a negative α_i value indicates the ith bit should be set to 1 to minimise the contribution from that term. For bivariate terms, a positive α_{ij} value indicates that the two bits x_i and x_j associated with it should be opposite in value, to minimise the contribution from the term involving $V(x_{ij})$. Similarly, a negative α_{ij} indicates that they should take the same value. This principle can be further extended to multivariate terms.

In [9] such analysis of MFM coefficients revealed a clear relationship with properties underlying the fitness function. This was despite having used a very small number (120) of function evaluations. Fitness was modelled for a bio-control problem, where bits set to 1 indicated times that nematode worms should be applied to a mushroom crop for control of the pest sciarid fly. The coefficients pointed towards application of the nematodes at points which matched the lifecycle of the fly larvae. Analysis of univariate and bivariate model coefficients for a number of further fitness functions was conducted in [4] and reinforced this finding.

The relationship between model coefficients and the global optimum is further illustrated here for the Checkerboard problem [2, 24]. The objective for this is to realise a grid with a checkerboard pattern of alternating 1s and 0s; each 1 should be surrounded by 0s and vice versa, not including corners. Interactions occur between

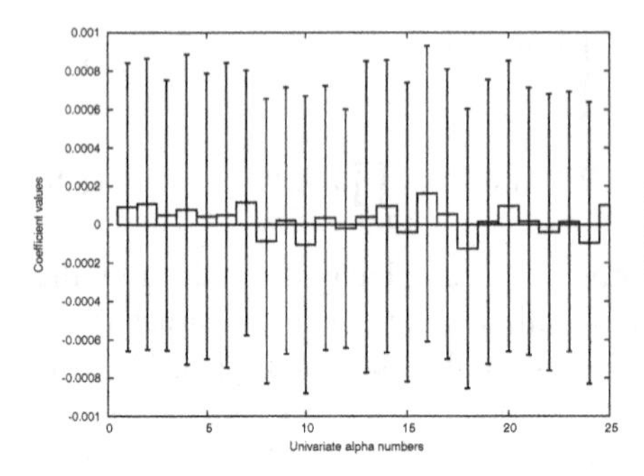

Fig. 8.1a Univariate alpha coefficients for 2D checkerboard

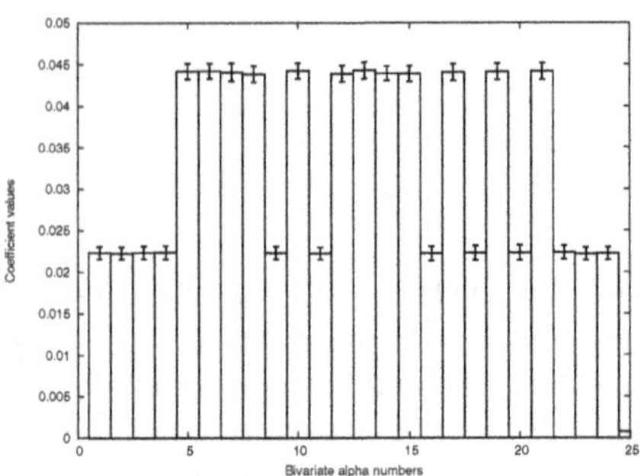

Fig. 8.1b Bivariate alpha coefficients for 2D checkerboard

neighbours on the lattice without wrapping around at the edges. The 2D lattice structure for a 25-bit instance of the problem is illustrated in Figure 8.2. We constructed the MFM using the perfect structure for the problem (that is, *univariate* terms for each variable X_i and *bivariate* terms for each neighbouring pair of variables X_iX_j on the lattice) and the model parameters were estimated using the fittest 220 solutions from a randomly generated starting population of 300 solutions. This process was repeated for 100 random starting populations, and the mean and standard deviation for each coefficient value over the 30 runs was computed. The coefficient values for the univariate and bivariate terms are illustrated in Figures 8.1a and 8.1b respectively. The univariate alphas are all close to zero: this is because there are two global optima which have complementary bits set to 1 and 0, so the model does not bias individual variables towards one value. The bivariate alphas are all positive, indicating that neighbouring variables should be opposite in value. Of particular interest is

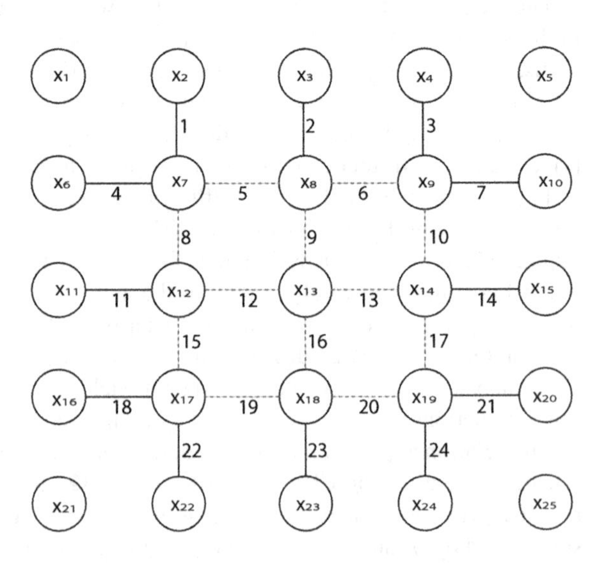

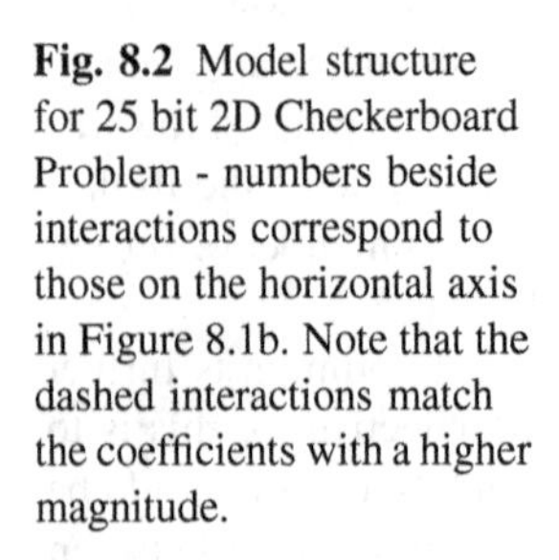

Fig. 8.2 Model structure for 25 bit 2D Checkerboard Problem - numbers beside interactions correspond to those on the horizontal axis in Figure 8.1b. Note that the dashed interactions match the coefficients with a higher magnitude.

Algorithm 1. Calculation of fitness prediction correlations

1: Generate random initial population p
2: Evaluate p using the fitness function and select a subset σ_1 of p
3: Use σ_1 to estimate MFM parameters
4: **for all** individuals in σ_1 **do**
5: Mutate one bit in the individual
6: Use MFM to predict fitness of individual
7: Use fitness function to determine true fitness
8: **end for**
9: Calculate the correlation coefficient between the predicted and true fitnesses (C_m)
10: Generate random population σ_2 equal in size to σ_1
11: **for all** individuals in σ_2 **do**
12: Use MFM to predict fitnesses
13: Use fitness function to determine true fitness
14: Calculate the correlation coefficient between the predicted and true fitnesses (C_r)
15: **end for**

that several of the bivariate alpha values are approximately double the magnitude of the others. The higher alpha values correspond to the interactions in the middle of the lattice (that is, neither of the variables they are associated with is on the edge of the checkerboard); these are dashed in Figure 8.2. These have a greater influence on fitness than those near the edge because if they break the constraint of neighbours not matching, their neighbours will also be affected. Thus we can see that the model places greater importance on these alphas. This shows that the MFM provides us with more information about the fitness function than simply pointing us in the direction of the global optimum.

A quantitative measure of model quality is also useful and the *Fitness prediction correlation* (FPC) [5] serves this purpose. This measures the MFM's ability to predict the fitness of unseen solutions. It is the Spearman's rank correlation [28] between the set of true fitnesses and fitnesses predicted by the model of an unseen population. Rank correlation is used because for discrete optimisation, it is only necessary to rank individuals in order of relative fitness. Predicting fitness is simply a reversal of the process used to estimate the α_K; the variable values for a solution being substituted into (8.9).

[7] defines two variants of FPC – C_m and C_r. C_r measures fitness prediction capability for randomly generated individuals; it follows that if this high then the MFM closely models the general fitness function. C_m measures fitness prediction capability for solutions neighbouring the current population, which is important for using the model to guide operators such as mutation. C_m and C_r are calculated by following Algorithm 1.

Given the right conditions (particularly when the model structure closely matches the non-zero components of the Walsh expansion of the fitness function, and large population), FPC values for the MFM can be close to 1 [4–7]. This indicates a strong correlation between predicted and true fitnesses for complex problems such as Ising and MAXSAT. This explains the success of the optimisation approaches using the MFM described in the next section.

8.3 Applications

There are several ways to exploit the model of fitness in the MFM. Within DEUM, direct sampling is used to generate new solutions with a high probability of being high in fitness [5, 53, 54]. This direct sampling of the fitness model rather than the fitness function has the benefit that the model can make the problem easier for the search part of the algorithm – the smoothing effect described in [63]. For example, in [54], it was found that directly sampling solutions from the MFM rather than the true fitness function for 2D Ising increased the success rate of a bitwise Gibbs sampler from 87% to 99%. It was concluded that by using a real-valued search space within the MFM rather than the discrete values within the fitness function, the fitness landscape was altered to allow the algorithm to escape being trapped within plateaux, leading to more efficient searching of the landscape.

8.3.1 MFM-GA

[8] proposes MFM-GA, applying the MFM as a surrogate in a genetic algorithm. The model is constructed at the start of the run, then the GA samples evaluations exclusively from the model. A number of benchmark functions were used as a proof-of-concept, then the algorithm was applied to a computationally expensive fitness function – feature selection for case-based reasoning. Promising results were reported, with a significant reduction in overall run-time over a GA. A major issue with the approach is that the model is only constructed once, making MFM-GA sensitive to modelling errors, with the result that the optimisation could not find solutions as fit as the GA. However, the solutions found were fitter than those found using a CBR-specific optimiser and represent something of a compromise between finding the fittest solution and a short run time.

8.3.2 Guided Local Search

A previously unpublished approach is to use the MFM as a fitness surrogate for a local search; allowing the local search many iterations without consuming many fitness evaluations. The MFM is used to filter solutions for evaluation by the true fitness function; this is similar to the approaches in [13, 25, 59].

This was applied with some success to the Huygens probe problem [29], part of a competition at the 2006 Congress on Evolutionary Computation (CEC). The objective of the problem is to find the lowest point on each of a series of 20 "moons" - fractal landscapes that are wrapped in both x and y dimensions. For each moon an algorithm is restricted to 1000 probes (fitness evaluations).

A bit-string encoding of the coordinate values was used. The algorithm rebuilds the model around progressively smaller areas of the lunar surface, suited to the fractal nature of the landscape (equal levels of detail at different zoom levels). Only univariate terms were included in the model; given that it is looking at neighbouring solutions to those used to build the model (those a short Hamming distance away), the results in Chapter 4 of [4] indicate that this should be enough to provide a reasonable fitness prediction capability. The workflow of the guided local search for this problem is given in Algorithm 2.

Algorithm 2. Guided Local Search

1: Generate random initial population p of size M
2: **while** chosen proportion of the available fitness evaluations not completely used **do**
3: Build univariate MFM modelling p
4: Truncation selection: select a subset σ, the fittest s individuals in p
5: **for all** individuals in σ **do**
6: Generate m neighbours μ:
7: Convert bit-string into real valued coordinates
8: Mutate values by up to a fixed amount (which decreases with each generation)
9: Convert numbers back to bit-string
10: **end for**
11: Use MFM to predict fitnesses of μ
12: Select predicted best l individuals ς from μ, calculate true fitnesses
13: Take best M from combined pool of p and ς and replace p
14: **end while**

Once the algorithm terminated with a single best solution x, the remainder of the 1000 evaluations were used for an exhaustive search of the neighbours to x. The proportion of evaluations allocated to each stage fixed per run; in the first instance the guided local search was given 2/3 of the total, in the second it was given 3/4.

Results for this algorithm were compared with the others taking part in the CEC competition. Each algorithm was run on the same set of 100 randomly generated moons, with a central server providing fitness evaluations and performing comparisons between algorithms via a SOAP interface. The algorithm performed comparably with a number of well-known problem solvers such as evolutionary strategies (ES), memetic algorithms (MA) and simulated annealing (SA), coming in 9th and 11th out of 16 with the two configurations. Unfortunately no more data is available on the specific implementations of these algorithms; however, these results do show that the approach is competitive with a wide range of others on this black-box problem.

8.4 Issues Affecting Modelling Capability

There are a number of factors which affect the quality of the fitness model, which also provide the grounding for future study.

8.4.1 Model Building Time

Of particular note is that the singular value decomposition used to estimate model parameters is $O(N^2 m)$ complexity in the number of model coefficients N and the population size m, meaning that building the model becomes expensive for increasing problem sizes and complexities. In much of our work, the population has had to be large enough to be slightly over-specified ($m > N$), so the overall model building complexity is $O(N^3)$. This is comparable to other EDAs – for example hBOA is dominated by Bayesian network structure learning complexity of $O(kn^2 m)$ [37], for problems of n variables which can be decomposed into subproblems of order k. Note however that for the MFM, the number of model coefficients N typically includes terms for each problem variable and for each interaction, so N is usually greater than the number of variables n in the problem, particularly for problems with many high-order interactions. A method of building the model incrementally or with multiple threads would help to mitigate this issue.

8.4.2 Model Structure

The structure of the MFM strongly influences its fitness modelling capability; in [4, 7] we observed the impact on fitness prediction capability of removing different cliques from the model. It is known that not all interactions which are present in a problem will necessarily be required in the model for the algorithm to rank individuals by fitness and find a global optimum. This observation is related to the concepts of necessary and unnecessary interactions [19] and benign and malign interactions [23]. This is also related to spurious correlations [45], false relationships in the model resulting from selection. Much of our previous work with the MFM has used a fixed structure derived from the problem definition, but for black-box problems in particular, the structure is unknown and must be inferred by sampling the fitness function. Works describing structure learning techniques for Markov networks include [43, 44, 46, 49]; the issue of structure learning specifically for the MFM in DEUM is explored in [6] and [51]. Approaches have typically involved conducting dependency tests such as Chi-Square on pairs of variables, either using existing members of the population or generating new solutions by mutating specific variables. A deterministic clique-finding algorithm can then be run on the resulting graph to find higher order cliques.

We further explore the issue of model and problem structure in relation to fitness modelling in [4, 7]. There we introduce the terms *perfect* and *imperfect*; perfect referring to the ideal structure with exactly the same interactions as present in the fitness function and imperfect referring any other structure.

8.4.3 Population Size and Selection

There is a clear and quantifiable relationship between the number of solutions present in the population used to estimate model parameters and the fitness mod-

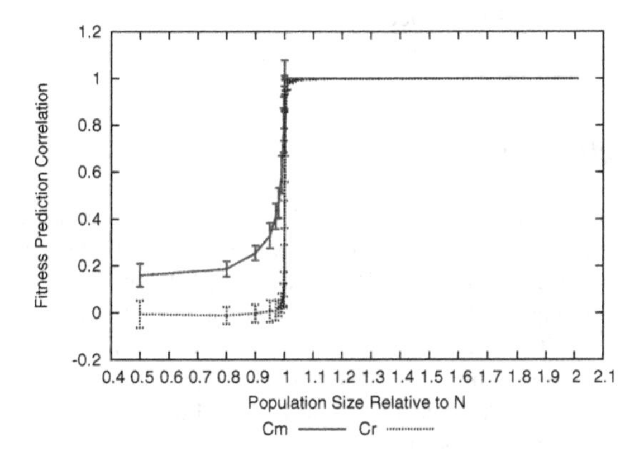

Fig. 8.3 FPC with increasing population size for 1000 bit onemax. Note the rapid increase in both FPC figures at the point where the population size exceeds 1001; the number of coefficients in the model for this problem (1000 univariate terms plus the constant).

elling capability. In [7] and more extensively in [4], a number of experiments show the effect of structure, population size and selection on the fitness prediction correlation. As population size reaches, then exceeds, the number of parameters in the model, there is a sharp transition from near zero to strong positive correlation between model and fitness function, for a number of fitness functions including onemax, Ising, MAXSAT and Trap-k. An example is given in Figure 8.3. With imperfect structures the model can typically predict fitnesses of solutions neighbouring those in the training population, but not those of randomly generated solutions.

The MFM is a probabilistic distribution over the fitness function, so sampling the model produces solutions with a high probability of being high in fitness. This means that explicit selection is not required for building the model; however selection still plays a crucial role in improving the fitness modelling capability [4, 7]. This is particularly the case where the population is too small or the model structure is imperfect, where selection sharpens the information about fitness already present in the population.

8.5 Conclusion and Future Work

In this chapter we have explained how the MFM applies Markov networks to fitness modelling of discrete problems. This may be exploited by directly sampling the model to generate new solutions, or by using the model as a surrogate, predicting fitness of solutions and filtering out promising ones for evaluation. Population size, model structure and selection are important factors in establishing a good model of fitness. Further exploration of these factors will help us to build better and more useful models of fitness, and also provide better understanding of how fitness is distributed within a population, which will be of use to the wider EA community. Further, the MFM is currently limited to fitness functions having a binary representation. Extension to higher cardinality or continuous representations would result in a more generally applicable model of fitness; this could entail adding additional terms to the model for each possible value. Finally, improving the efficiency of the model building process will increase the number of situations where it offers an overall performance improvement.

References

1. Abboud, K., Schoenauer, M.: Surrogate Deterministic Mutation: Preliminary Results. In: Collet, P., Fonlupt, C., Hao, J.-K., Lutton, E., Schoenauer, M. (eds.) EA 2001. LNCS, vol. 2310, pp. 104–116. Springer, Heidelberg (2002)
2. Baluja, S., Davies, S.: Using optimal dependency-trees for combinational optimization. In: ICML 1997: Proceedings of the Fourteenth International Conference on Machine Learning, pp. 30–38. Morgan Kaufmann Publishers Inc. (1997)
3. Bethke, A.: Genetic Algorithms as Function Optimizers. Ph.D. thesis, University of Mitchigan (1980)
4. Brownlee, A.E.I.: Multivariate Markov Networks for Fitness Modelling in an Estimation of Distribution Algorithm. Ph.D. thesis, Robert Gordon University, Aberdeen (2009), `http://hdl.handle.net/10059/381`
5. Brownlee, A.E.I., McCall, J.A.W., Brown, D.F.: Solving the MAXSAT problem using a multivariate EDA based on Markov networks. In: Proceedings of the Genetic and Evolutionary Computation Conference (GECCO 2007) (Late Breaking Papers), pp. 2423–2428. ACM Press, New York (2007)
6. Brownlee, A.E.I., McCall, J.A.W., Shakya, S.K., Zhang, Q.: Structure Learning and Optimisation in a Markov-network based Estimation of Distribution Algorithm. In: Proceedings of the IEEE Congress on Evolutionary Computation (CEC 2009), pp. 447–454. IEEE Press, Trondheim (2009)
7. Brownlee, A.E.I., McCall, J.A.W., Zhang, Q., Brown, D.: Approaches to Selection and their effect on Fitness Modeling in an Estimation of Distribution Algorithm. In: Proceedings of the IEEE World Congress on Computational Intelligence (CEC 2008), pp. 2621–2628. IEEE Press, Hong Kong (2008)
8. Brownlee, A.E.I., Regnier-Coudert, O., McCall, J.A.W., Massie, S.: Using a Markov network as a surrogate fitness function in a genetic algorithm. In: Proceedings of the IEEE Congress on Evolutionary Computation (CEC 2010), pp. 4525–4532. IEEE Press, Barcelona (2010)
9. Brownlee, A.E.I., Wu, Y., McCall, J.A.W., Godley, P.M., Cairns, D.E., Cowie, J.: Optimisation and fitness modelling of bio-control in mushroom farming using a Markov network EDA. In: Proceedings of the Genetic and Evolutionary Computation Conference (GECCO 2008), pp. 465–466. ACM, Atlanta (2008)
10. Bui, L.T., Abbass, H.A., Essam, D.: Fitness inheritance for noisy evolutionary multi-objective optimization. In: Proceedings of the 2005 Conference on Genetic and Evolutionary Computation, GECCO 2005, pp. 779–785. ACM, New York (2005)
11. Chen, J.H., Goldberg, D., Ho, S.-Y., Sastry, K.: Fitness inheritance in multiobjective optimization. In: Proceedings of the Genetic and Evolutionary Computation COnference (GECCO 2002), pp. 319–326. ACM Press (2002)
12. Davarynejad, M., Ahn, C.W., Vrancken, J., van den Berg, J., Coello Coello, C.A.: Evolutionary hidden information detection by granulation-based fitness approximation. Appl. Soft Comput. 10, 719–729 (2010)
13. Emmerich, M., Giannakoglou, K., Naujoks, B.: Single- and multiobjective evolutionary optimization assisted by gaussian random field metamodels. IEEE Transactions on Evolutionary Computation 10(4), 421–439 (2006)

14. Furtuna, R., Curteanu, S., Leon, F.: An elitist non-dominated sorting genetic algorithm enhanced with a neural network applied to the multi-objective optimization of a polysiloxane synthesis process. Eng. Appl. Artif. Intell. 24, 772–785 (2011)
15. Goldberg, D.: Genetic Algorithms and Walsh Functions: Part I, A Gentle Introduction. Complex Systems 3, 129–152 (1989)
16. Goldberg, D.: Genetic Algorithms and Walsh Functions: Part II, Deception and its Analysis. Complex Systems 3, 153–171 (1989)
17. Golubov, B., Efimov, A., Skvortsov, V.: Walsh Series and Transforms: Theory and Applications. Mathematics and Applications: Soviet Series, vol. 64. Kluwer Academic Publishers, Boston (1991)
18. Han, S.H., Yang, H.: Screening important design variables for building a usability model: genetic algorithm-based partial least-squares approach. International Journal of Industrial Ergonomics 33(2), 159–171 (2004)
19. Hauschild, M., Pelikan, M., Lima, C.F., Sastry, K.: Analyzing probabilistic models in hierarchical BOA on traps and spin glasses. In: Proceedings of the Genetic and Evolutionary Computation Conference (GECCO 2007), pp. 523–530. ACM Press, New York (2007)
20. Jin, Y.: A comprehensive survey of fitness approximation in evolutionary computation. Soft Computing 9(1), 3–12 (2005)
21. Jin, Y., Sendhoff, B.: Reducing Fitness Evaluations Using Clustering Techniques and Neural Network Ensembles. In: Deb, K., et al. (eds.) GECCO 2004, Part I. LNCS, vol. 3102, pp. 688–699. Springer, Heidelberg (2004)
22. Johanson, B.: Gp-music: An interactive genetic programming system for music generation with automated fitness raters. In: Proceedings of the Third Annual Conference, pp. 181–186. MIT Press (1998)
23. Kallel, L., Naudts, B., Reeves, R.: Properties of fitness functions and search landscapes. In: Kallel, L., Naudts, B., Rogers, A. (eds.) Theoretical Aspects of Evolutionary Computing, pp. 177–208. Springer (2000)
24. Larrañaga, P., Lozano, J.A.: Estimation of Distribution Algorithms: A New Tool for Evolutionary Computation. Kluwer Academic Publishers, Boston (2002)
25. Li, R., Emmerich, M., Eggermont, J., Bovenkamp, E., Back, T., Dijkstra, J., Reiber, J.: Metamodel-assisted mixed integer evolution strategies and their application to intravascular ultrasound image analysis. In: IEEE Congress on Evolutionary Computation, CEC 2008 (IEEE World Congress on Computational Intelligence), pp. 2764–2771 (2008)
26. Lim, D., Jin, Y., Ong, Y.S., Sendhoff, B.: Generalizing surrogate-assisted evolutionary computation. IEEE Transactions on Evolutionary Computation 14(3), 329–355 (2010)
27. Lima, C.F., Sastry, K., Goldberg, D.E., Lobo, F.G.: Combining competent crossover and mutation operators: a probabilistic model building approach. In: Proceedings of the 2005 Conference on Genetic and Evolutionary Computation (GECCO 2005), pp. 735–742. ACM, New York (2005)
28. Lucey, T.: Quantatitive Techniques: An Instructional Manual. D. P. Publications, Eastleigh (1984)
29. MacNish, C.: Benchmarking Evolutionary Algorithms: The Huygens Suite. In: Proceedings of the Genetic and Evolutionary Computation Conference (GECCO 2005) (Late Breaking Papers), pp. 2423–2428. ACM Press, New York (2005)
30. Macready, W., Levitan, B.: Learning landscapes: regression on discrete spaces. In: Proceedings of the IEEE Congress on Evolutionary Computation (CEC 1999), vol. 1, pp. 687–694 (1999)

31. Magnier, L., Haghighat, F.: Multiobjective optimization of building design using trnsys simulations, genetic algorithm, and artificial neural network. Building and Environment 45(3), 739–746 (2010)
32. Michalski, R.S.: Learnable evolution model: Evolutionary processes guided by machine learning. Machine Learning 38(1-2), 9–40 (2000)
33. Miquélez, T., Bengoetxea, E., Larrañaga, P.: Evolutionary computation based on Bayesian classifiers. International Journal of Applied Mathematics and Computer Science 14(3), 101–115 (2004)
34. Ochoa, A.: Opportunities for Expensive Optimization with Estimation of Distribution Algorithms. In: Tenne, Y., Goh, C.-K. (eds.) Computational Intel. in Expensive Opti. Prob. ALO, vol. 2, pp. 193–218. Springer, Heidelberg (2010)
35. Ochoa, A.A., Soto, M.R.: Partial evaluation in genetic algorithms. In: Proceedings of the 10th International Conference on Industrial and Engineering Applications of Artificial Intelligence and Expert Systems, IEA/AIE 1997, pp. 217–222. Goose Pond Press (1997)
36. Peña, J.M., Robles, V., Larrañaga, P., Herves, V., Rosales, F., Pérez, M.S.: GA-EDA: Hybrid Evolutionary Algorithm Using Genetic and Estimation of Distribution Algorithms. In: Orchard, B., Yang, C., Ali, M. (eds.) IEA/AIE 2004. LNCS (LNAI), vol. 3029, pp. 361–371. Springer, Heidelberg (2004)
37. Pelikan, M.: Hierarchical Bayesian Optimization Algorithm - Toward a New Generation of Evolutionary Algorithms. STUDFUZZ, vol. 170. Springer (2005)
38. Pelikan, M., Sastry, K.: Fitness Inheritance in the Bayesian Optimization Algorithm. In: Deb, K., et al. (eds.) GECCO 2004, Part II. LNCS, vol. 3103, pp. 48–59. Springer, Heidelberg (2004)
39. Press, W.H., Flannery, B.P., Teukolsky, S.A., Vetterling, W.T.: Numerical Recipes: The Art of Scientific Computing. Cambridge University Press, Cambridge (1986)
40. Rasheed, K., Vattam, S., Ni, X.: Comparison of methods for using reduced models to speed up design optimization. In: Proceedings of the Genetic and Evolutionary Computation COnference (GECCO 2002), pp. 1180–1187. Morgan Kaufmann Publishers Inc., San Francisco (2002)
41. Regis, R., Shoemaker, C.: Local function approximation in evolutionary algorithms for the optimization of costly functions. IEEE Transactions on Evolutionary Computation 8(5), 490–505 (2004)
42. Sano, Y., Kita, H.: Optimization of noisy fitness functions by means of genetic algorithms using history of search with test of estimation. In: Proceedings of the World on Congress on Computational Intelligence, vol. 1, pp. 360–365 (2002)
43. Santana, R.: A Markov Network Based Factorized Distribution Algorithm for Optimization. In: Lavrač, N., Gamberger, D., Todorovski, L., Blockeel, H. (eds.) ECML 2003. LNCS (LNAI), vol. 2837, pp. 337–348. Springer, Heidelberg (2003)
44. Santana, R., Larrañaga, P., Lozano, J.A.: Mixtures of Kikuchi Approximations. In: Fürnkranz, J., Scheffer, T., Spiliopoulou, M. (eds.) ECML 2006. LNCS (LNAI), vol. 4212, pp. 365–376. Springer, Heidelberg (2006)
45. Santana, R., Larrañaga, P., Lozano, J.A.: Research topics on discrete estimation of distribution algorithms. Memetic Computing 1(1), 35–54 (2009)
46. Santana, R., Ochoa, A., Soto, M.R.: The mixture of trees factorized distribution algorithm. In: Spector, L., Goodman, E., Wu, A., Langdon, W.B., Voigt, H.M., Gen, M., Sen, S., Dorigo, M., Pezeshk, S., Garzon, M., Burke, E. (eds.) Proceedings of the Genetic and Evolutionary Computation COnference (GECCO 2001), pp. 543–550. Morgan Kaufmann Publishers (2001)

47. Sastry, K., Lima, C., Goldberg, D.E.: Evaluation relaxation using substructural information and linear estimation. In: Proceedings of the Genetic and Evolutionary Computation COnference (GECCO 2006), pp. 419–426. ACM Press, New York (2006)
48. Schmidt, M.D., Lipson, H.: Coevolution of fitness predictors. IEEE Transactions on Evolutionary Computation 12(6), 736–749 (2008)
49. Shakya, S., Santana, R.: A Markovianity based optimisation algorithm. Tech. Rep. EHU-KZAA-IK-3/08, Department of Computer Science and Artificial Intelligence, University of the Basque Country (2008)
50. Shakya, S.K.: DEUM: A framework for an estimation of distribution algorithm based on Markov random fields. Ph.D. thesis, The Robert Gordon University, Aberdeen, UK (2006), http://hdl.handle.net/10059/39
51. Shakya, S.K., Brownlee, A.E.I., McCall, J.A.W., Fournier, F., Owusu, G.: A fully multivariate DEUM algorithm. In: Proceedings of the IEEE Congress on Evolutionary Computation (CEC 2009), pp. 479–486. IEEE Press (2009)
52. Shakya, S.K., McCall, J.A.W.: Optimization by estimation of distribution with DEUM framework based on Markov random fields. International Journal of Automation and Computing 4(3), 262–272 (2007)
53. Shakya, S.K., McCall, J.A.W., Brown, D.F.: Incorporating a Metropolis method in a distribution estimation using Markov random field algorithm. In: Proceedings of the IEEE Congress on Evolutionary Computation (CEC 2005), pp. 2576–2583. IEEE Press (2005)
54. Shakya, S.K., McCall, J.A.W., Brown, D.F.: Solving the Ising spin glass problem using a bivariate EDA based on Markov random fields. In: Proceedings of the IEEE World Congress on Computational Intelligence (CEC 2006). IEEE Press (2006)
55. Shi, L., Rasheed, K.: A Survey of Fitness Approximation Methods Applied in Evolutionary Algorithms. In: Tenne, Y., Goh, C.K. (eds.) Computational Intelligence in Expensive Optimization Problems. Adaptation Learning and Optimization, vol. 2, ch.1, pp. 3–28. Springer, Heidelberg (2010)
56. Smith, R.E., Dike, B.A., Stegmann, S.A.: Fitness inheritance in genetic algorithms. In: SAC 1995: Proceedings of the 1995 ACM Symposium on Applied Computing, pp. 345–350. ACM Press, New York (1995)
57. Syberfeldt, A., Grimm, H., Ng, A., John, R.I.: A parallel surrogate-assisted multi-objective evolutionary algorithm for computationally expensive optimization problems. In: Wang, J. (ed.) Proceedings of the IEEE World Congress on Computational Intelligence (CEC 2008), pp. 3177–3184. IEEE Computational Intelligence Society, IEEE Press, Hong Kong (2008)
58. Takahashi, S., Kita, H., Suzuki, H., Sudo, T., Markon, S.: Simulation-based optimization of a controller for multi-car elevators using a genetic algorithm for noisy fitness function. In: Proceedings of the IEEE Congress on Evolutionary Computation (CEC 2003), vol. 3, pp. 1582–1587 (2003)
59. Wallin, D., Ryan, C.: Using over-sampling in a Bayesian classifier EDA to solve deceptive and hierarchical problems. In: Proceedings of the IEEE Congress on Evolutionary Computation (CEC 2009), pp. 1660–1667 (2009)
60. Zhang, Q., Sun, J.: Iterated local search with guided mutation. In: Proceedings of the IEEE World Congress on Computational Intelligence (CEC 2006), pp. 924–929. IEEE Press (2006)
61. Zhang, Q., Sun, J., Tsang, E.: Combinations of estimation of distribution algorithms and other techniques. International Journal of Automation & Computing, 273–280 (2007)

62. Zhou, L., Haghighat, F.: Optimization of ventilation system design and operation in office environment, part i: Methodology. Building and Environment 44(4), 651–656 (2009)
63. Zhou, Z., Ong, Y.S., Lim, M.H., Lee, B.S.: Memetic algorithm using multi-surrogates for computationally expensive optimization problems. Soft Comput. 11(10), 957–971 (2007)
64. Zhou, Z., Ong, Y.S., Nguyen, M.H., Lim, D.: A study on polynomial regression and Gaussian process global surrogate model in hierarchical surrogate-assisted evolutionary algorithm. In: Proceedings of the IEEE Congress on Evolutionary Computation (CEC 2005), vol. 3, pp. 2832–2839 (2005)
65. Ziegler, J., Banzhaf, W.: Decreasing the Number of Evaluations in Evolutionary Algorithms by Using a Meta-Model of the Fitness Function. In: Ryan, C., Soule, T., Keijzer, M., Tsang, E.P.K., Poli, R., Costa, E. (eds.) EuroGP 2003. LNCS, vol. 2610, pp. 264–275. Springer, Heidelberg (2003)

Chapter 9
Fast Fitness Improvements in Estimation of Distribution Algorithms Using Belief Propagation

Alexander Mendiburu, Roberto Santana, and Jose A. Lozano

Abstract. Factor graphs can serve to represent Markov networks and Bayesian networks models. They can also be employed to implement efficient inference procedures such as belief propagation. In this paper we introduce a flexible implementation of belief propagation on factor graphs in the context of estimation of distribution algorithms (EDAs). By using a transformation from Bayesian networks to factor graphs, we show the way in which belief propagation can be inserted within the Estimation of Bayesian Networks Algorithm (EBNA). The objective of the proposed variation is to increase the search capabilities by extracting information of the, computationally costly to learn, Bayesian network. Belief Propagation applied to graphs with cycles allows to find (with a low computational cost), in many scenarios, the point with the highest probability of a Bayesian network. We carry out some experiments to show how this modification can increase the potentialities of Estimation of Distribution Algorithms.

9.1 Introduction

In the last ten years Estimation of distribution algorithms (EDAs) [18, 28, 34] have turned in a lively research area inside the evolutionary computation field. Like most evolutionary computation heuristics, EDAs maintain at each step a population of individuals but, instead of using crossover and mutation operators to make the population evolve, they learn a probability distribution from the set of selected individuals, sampling this distribution to obtain the new population. A key point in these algorithms is the way estimation of the joint probability distribution is done. According to this, three groups could be differentiated: algorithms that assume that all the variables are independent, those that assume second order statistics, and those that consider unrestricted models.

Alexander Mendiburu · Roberto Santana · Jose A. Lozano
Intelligent Systems Group, The University of the Basque Country (UPV/EHU)
e-mail: {alexander.mendiburu, roberto.santana, ja.lozano}@ehu.es

S. Shakya and R. Santana (Eds.): Markov Networks in Evolutionary Computation, ALO 14, pp. 141–155.
springerlink.com

In the field of combinatorial optimization, the algorithms that use unrestricted models (usually by means of Bayesian networks) are the most commonly used [17, 27, 33]). Although the algorithms using Bayesian networks get the best results in terms of objective function value, they suffer from the cost of the learning of the probabilistic graphical model at each step. In fact, learning a Bayesian network is an NP-hard problem[3] and therefore local (in general heuristic) algorithms need to be used at each step. However, even with the use of these local search algorithms or even approaches such as updating the model every few generations (instead of doing it at each generation) [35], the computational cost implied by learning dominates in most of the cases the algorithm runtime (putting aside the cost of evaluating the objective function). Due to this fact, researchers have introduced different methods and techniques trying to take advantage of the learnt graphical model, improving this way the behavior and performance of EDAs. This can be done in different forms: The graphical model can be used as a source of information for sophisticated local optimizers [20, 36, 43], the graphical model can serve for implementing partial or full estimation of the fitness function [30, 44–46] or be employed as a model of the process being optimized [41]. In all these cases, an important issue is the way in which the information contained in the probabilistic model is extracted. This question is relevant for the sampling step of EDAs, and can be crucial for the efficient implementation of the techniques previously mentioned.

In this chapter, we focus on a step that has not received much attention from the researchers: the problem of efficiently sampling probabilistic graphical models in EDAs. In fact, we investigate the effect of explicitly adding the point of highest probability in the graphical model to the population. To compute this point, known as the most probable configuration of the model, we use belief propagation (BP) algorithms [32, 50, 51]. These algorithms are commonly used in graphical models to carry out inference tasks. For instance, they can be used to reason in probabilistic graphical models (calculate marginal probabilities or posterior probabilities), or to calculate the point with the highest probability. Recently, blending of BP algorithms with techniques coming from the field of statistical physics, particularly algorithms developed to minimize the free energy [51], has brought new ideas to the field. One of these ideas is related to the application of BP algorithms in graphs with cycles (in this case the algorithm is usually called loopy belief propagation but for the sake of simplicity we will use through the paper the most general BP term). While the convenient convergence and exactness properties of BP in acyclic graphs are lost, it has been widely proved in many applications that these algorithms often produce the expected results with a low computational cost [4, 5, 50].

We present, therefore, an exploratory work that studies disregarded aspects of EDAs, incorporating techniques that are widely and successfully applied in other fields. This paper extends a preliminary work [23] including more problems and instances in order to obtain sound conclusions. However, it is important to note that the work presented in this chapter is just one example of a variety of possible applications of BP within EDAs. The scope of application of belief propagation algorithms in EDAs is very wide. It covers EDAs that use directed, undirected and more sophisticated graphical models. For instance, BP could be applied to undirected probabilistic graphical models that are directly learned from data and not the product of a transformation as the case presented here. BP and other variants of belief propagation algorithms could also be applied to combine a priori knowledge about the problem structure, possibly represented by an undirected graph, with information captured by the probabilistic model. Comparing the use of BP to previous proposals, while methods for computing the most probable configuration produce global information about the network, the use of substructural neighborhood, as done in [44], can be considered as a way to use local information of the graphical model. Another difference is related to the aims of the methods. Substructural neighborhoods are used for local optimization, and most probable configuration algorithms are general inference methods. Fitness estimation can provide information about the quality of any solution, but in general it can not be used as a procedure to determine the solution with the best estimation given the fitness model, as methods for finding the most probable configuration do. Other algorithms [15] that combine message passing with accurate approximation methods from statistical physics can be used for finding the most probable configurations and provide good approximations of the fitness. Local optimization techniques based on graphical models and general fitness estimation procedures can be used to enhance the algorithms that find the most probable configurations. Finally, there are previous reports on the application of BP algorithms in EDAs. In [31], BP is applied to improve the results achieved by the polytree distribution algorithm (PADA). The objective is to construct higher-order marginal distributions from the bivariate marginals corresponding to a polytree distribution. These results are extended in [14] to the factorized distribution algorithm (FDA) with fixed structure. In contrast to the previous approaches, our proposal allows the application of the BP algorithm on more complex models than those learned by PADA, and the structure is not fixed from the beginning as is the case of FDA. In [42] affinity-propagation (another message-passing technique) is used for learning the structure of the probabilistic model.

In [47], an algorithm that calculates the most probable configurations in the context of optimization by EDAs was introduced. The algorithm was applied to obtain the most probable configurations of the univariate marginal model used by the univariate marginal distribution algorithm [25], and models based on trees and polytrees. These results were extended in [39] to pairwise Markov networks which are covered by the more general factor graphs we used in our proposal. Recently, more sophisticated BP algorithms have been used in the general context of optimization based on probabilistic models for obtaining higher order consistent marginal probabilities [26], as well as the most probable configurations of the model [15]. Also in these cases, the structure of the problem is known a priori. Finally, based on our initial work, Lima et al. [19] use BP to perform a substructural local search, looking for the solution with the highest fitness value instead of that with the highest probability.

Our proposal rests in the general schema of EDAs and it is more general than those previously presented as it allows the use of non-fixed unrestricted graphical models. Taking into account that there are several implementations of EDAs that use Bayesian networks, we consider that the generality of our approach allows it to be inserted in such implementations.

The remainder of the paper is organized as follows. In Section 9.2 the main characteristics of the BP algorithm will be introduced. Section 9.3 explains the new EDA+BP proposal and Section 9.4 presents the experiments completed to observe the influence of BP in EDAs. Finally, Section 9.5 concludes the paper.

9.2 Belief Propagation

BP [32] is a widely recognized method to solve inference problems in graphical models. It is mainly applied to two different situations: (1) when the goal is to obtain marginal and/or posterior probabilities for some of the variables in the problem, and (2) with the aim of searching for the most probable global state of a problem given its model. These two variants are also known as the *sum-product* and *max-product* algorithms.

The BP algorithm has been proved to be efficient on tree-shaped structures, and empirical experiments have often shown good approximate outcomes even when applied to cyclic graphs. This has been widely demonstrated in many applications including low-density parity-check codes [38], turbo codes [21], image processing [10], and optimization [2]. We illustrate BP by means of a probabilistic graphical model called factor graph. Factor graphs are bipartite graphs with two different types of nodes: variable nodes and factor nodes. Each variable node identifies a single variable X_i that can take values from a discrete domain, while factor nodes f_j represent different functions whose arguments are subsets of variables. This is graphically represented by edges that connect a particular function node with its variable nodes (arguments). Figure 9.1 shows a simple factor graph with six variable nodes $\{X_1, X_2, \ldots, X_6\}$ and three factor nodes $\{f_a, f_b, f_c\}$.

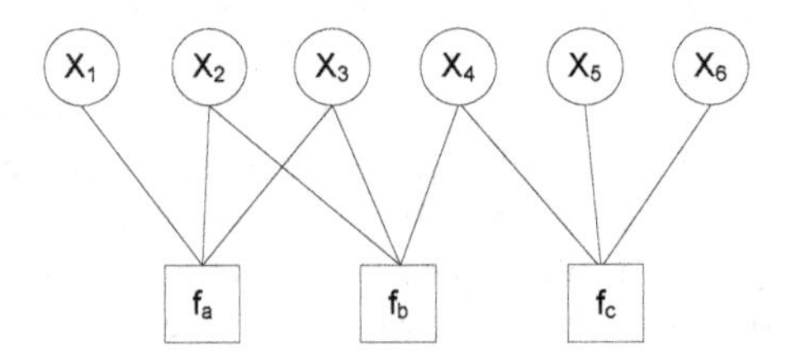

Fig. 9.1 Example of a factor graph.

Factor graphs are appropriate to represent those cases in which the joint probability distribution can be expressed as a factorization of several local functions:

$$g(x_1,\ldots,x_n) = \frac{1}{Z}\prod_{j\varepsilon J} f_j(x_j) \tag{9.1}$$

where $Z = \sum_x \prod_{j\varepsilon J} f_j(x_j)$ is a normalization constant, n is the number of variable nodes, J is a discrete index set, X_j is a subset of $\{X_1,\ldots,X_n\}$, and $f_j(x_j)$ is a function containing the variables of X_j as arguments. Applying this factorization to the factor graph presented in Figure 9.1, the joint probability distribution would result in:

$$g(x_1,\ldots,x_6) = \frac{1}{Z} f_a(x_1,x_2,x_3) f_b(x_2,x_3,x_4) f_c(x_4,x_5,x_6) \tag{9.2}$$

The main characteristic of the BP algorithm is that the inference is done using message-passing between nodes. Each node sends and receives messages until a stable situation is reached. Messages, locally calculated by each node, comprise statistical information concerning neighbor nodes.

When using BP with factor graphs, two kinds of messages are identified: messages $n_{i\to a}(x_i)$ sent from a variable node i to a factor node a, and messages $m_{a\to i}(x_i)$ sent from a factor node a to a variable node i. Note that a message is sent for every value of each variable X_i.

These messages are updated according to the following rules:

$$n_{i\to a}(x_i) := \prod_{c \varepsilon N(i)\setminus a} m_{c\to i}(x_i) \tag{9.3}$$

$$m_{a\to i}(x_i) := \sum_{x_a \setminus x_i} f_a(x_a) \prod_{j \varepsilon N(a)\setminus i} n_{j\to a}(x_j) \tag{9.4}$$

$$m_{a\to i}(x_i) := \arg\max_{x_a \setminus x_i} \{ f_a(x_a) \prod_{j \varepsilon N(a)\setminus i} n_{j\to a}(x_j) \} \tag{9.5}$$

where $N(i)\setminus a$ represents all the neighboring factor nodes of node i excluding node a, and $\sum_{x_a \setminus x_i}$ expresses that the sum is completed taking into account all the possible values that all variables but X_i in X_a can take –while variable X_i takes its x_i value.

Equations 9.3 and 9.4 are used when marginal probabilities are looked for (sum-product). By contrast, in order to obtain the most probable configurations (max-product), Equations 9.3 and 9.5 should be applied.

When the algorithm converges (i.e. messages do not change), marginal functions (sum-product) or max-marginals (max-product) are obtained as the normalized product of all messages received by X_i:

$$g_i(x_i) \propto \prod_{a \varepsilon N(i)} m_{a\to i}(x_i) \tag{9.6}$$

Regarding the max-product approach, when the algorithm converges to the most probable value, each variable in the optimal solution is assigned the value given by the configuration with the highest probability at each max-marginal. Some theoretical results on BP and modifications for maximization can be found in [49].

As described previously, BP is a widely studied and used algorithm, and has been rediscovered and adapted repeatedly to particular problems. Thus, different implementations have been developed since the algorithm was first proposed. As this algorithm uses message-passing to complete inference, the way and the order in which messages are sent plays a crucial role. In addition, other aspects such as the initial values or the stopping criteria can lead the algorithm to different results.

For this work, we have used a BP implementation introduced in [24]. This is a flexible and distributed version of the BP algorithm providing different configuration parameters to the final user, such as message scheduling policies –i.e. when and how the messages are sent and received–, stopping criteria and initial values for some parameters –including for example, the initial values of the messages.

9.3 EDAs and BP

Among the wide range of algorithms that belong to the family of EDAs, in this paper we focus on estimation of Bayesian networks algorithm (EBNA) [17]. This algorithm learns and samples a Bayesian network at each step. However, our proposal could be easily applied to any other EDA such as the Bayesian optimization algorithm (BOA) [33], or the Learning Factorized Distribution Algorithm (LFDA) [27].

The most costly step in EBNA (apart from the computation of the objective function) is the learning of the Bayesian network (this is usually done by means of a local search algorithm that adds or removes at each step the arc that increases a given score the most). The sampling step is commonly done by means of Probabilistic Logic Sampling (PLS) [13] where the instantiation of the variables is done starting from the root and following an ancestral ordering determined by the Bayesian network structure.

PLS is a very easy to implement, low complexity algorithm but it has an important limitation: If the probability associated to the most probable solutions (which in the context of EDAs are expected to match, at least approximately, the fittest individuals) is too small, then PLS will likely miss these points. This means that we are not taking advantage to a full extent of the information contained in the model. So, we propose to apply BP to obtain the point or points with the highest probability and combine them with those points generated by PLS.

To do that, first PLS will be used (as usual) to sample individuals from the Bayesian network. Then, BP will be applied to obtain an additional individual. Then, the individual obtained by using BP will be incorporated into the population providing that it is better (in terms of fitness) than the best individual of the population. In order to use BP, first we turn the Bayesian network into a factor graph by assigning a factor node to each variable and its parent set (a variable node is created for each variable of the problem) [16]. Then, BP is applied in the factor graph to obtain the new individual.

Regarding the computational cost, it must be remained that BP is exponentially complex in the number of variables in the cliques. Each message takes $O(k^M)$ where k is the number of variables in the clique and M is the number of states. Therefore, BP is particularly slow for large cliques. The algorithm maybe especially slow for continuous variables. Consequently, loopy BP over continuous variables is usually limited to pairwise MRFs. Nevertheless, some alternatives can be used to diminish the complexity of the algorithm. One possibility is the use of parallel algorithms [24]. Another alternative is the introduction of linear constraint nodes [37].

9.4 Experiments

In order to test the benefits that incorporating BP can bring, we completed several experiments to compare the behavior of EBNA and EBNA with BP (from now on EBNA-BP). For this purpose, we have chosen five different problems: *OneMax*, *CheckerBoard*, *±J Ising spin glass*, *Gaussian Ising spin glass*, and *Trap5*.

OneMax is a well-known toy function which has a very simple objective function. This problem consists of maximizing:

$$OneMax(x) = \sum_{i=1}^{n} x_i \tag{9.7}$$

where $x_i \in \{0,1\}$.

The *Checkerboard* function was first introduced in [1]. In this problem, a $s \times s$ grid is given. Each point of the grid can take a value 0 or 1. The goal of the problem is to create a checkerboard pattern of 0's and 1's on the grid. Each point with a value of 1 should be surrounded in all four basic directions by a value of 0, and vice versa. The evaluation counts the number of correct surrounding bits. The maximum value is $2s(s-1)$, and the problem dimension is $n = s^2$. If we consider the grid as a matrix $x = [x_{ij}]_{i,j=1,\dots,s}$ and interpret $\delta(a,b)$ as the Kronecker's delta function, the checkerboard function can be written as[1]:

$$F_{CheckerBoard}(x) = 2s(s-1) - \sum_{i=1}^{s}\sum_{j=1}^{s-1} \delta(x_{ij}, x_{ij+1}) - \sum_{j=1}^{s}\sum_{i=1}^{s-1} \delta(x_{ij}, x_{i+1j})$$

Ising spin glass is an optimization problem which has been solved and analyzed in different works related with EDAs [12]. A classic *2D Ising spin glass* can be simply formulated. The set of variables X is seen as a set of n spins disposed on a regular 2D grid L with $n = ll$ sites and periodic boundaries. Each node of L corresponds to a spin X_i and each edge (i, j) corresponds to a coupling between X_i and X_j. Thus, each spin variable interacts with its four nearest neighbors in the toroidal structure L. Moreover, each edge of L has an associated coupling strength J_{ij} between the related spins. The target is, given couplings J_{ij}, to find the spin configuration that minimizes the energy of the system computed as,

$$E(x) = -\sum_{i<j \in L} J_{ij} x_i x_j - \sum_{i \in L} h_i x_i \tag{9.8}$$

where the sum runs over all coupled spins. In our experiments we take $h_i = 0 \, \forall i \in L$. The states with minimum energy are called ground states. Depending on the range

[1] This is a slightly different formulation from the one used in [1].

chosen for the couplings J_{ij} we have different versions of the problem. For the *Gaussian Ising* problem, the couplings J_{ij} are real numbers generated following a Gaussian distribution. On the other hand, for the $\pm J$ *Ising* problem, the couplings J_{ij} are $+1$ or -1.

Trap5 [6], is an additively separable (non overlapping) function with a unique optimum. It divides the set X of n variables, into disjoint subsets X_I of 5 variables. It can be defined using a unitation function $u(y) = \sum_{i=1}^{p} y_i$ where $y \in \{0,1\}^p$ as,

$$\text{Trap5}(x) = \sum_{I=1}^{\frac{n}{5}} trap_5(x_I) \tag{9.9}$$

where $trap_5$ is defined as,

$$trap_5(x_I) = \begin{cases} 5 & \text{if } u(x_I) = 5 \\ 4 - u(x_I) & \text{otherwise} \end{cases} \tag{9.10}$$

and $x_I = (x_{5I-4}, x_{5I-3}, x_{5I-2}, x_{5I-1}, x_{5I})$ is an assignment to each trap partition X_I. This function has one global optimum in the assignment of all ones for X and a large number of local optima, $2^{n/5} - 1$.

The *Trap5* function has been used in previous works to study the structure of the probabilistic models in EDAs based on Bayesian networks as well as studying the influence of different parameters [12]. It is important to note that this function is difficult to optimize if the probabilistic model is not able to identify interactions between variables [8]. Therefore, we have included an additional experiment using a fixed complete structure for Trap5, which guarantees a fit probabilistic model.

9.4.1 Experiments Settings

EBNA and EBNA-BP were run under the same conditions and using the same parameters for the five problems: Individual size ($n = 100$), population size ($N = 2,000$), random generation of the initial population, truncation selection ($T = 0.5$), elitism (the best individual of the previous generation is guaranteed to survive), termination criterion set based on a maximum number of generations (100). For the BP algorithm, a set-based scheduling policy was used (from all to all neighbors), allowed difference is 10^{-3}, the maximum number of messages are 1000, cache-size is 100, and 10 is the number of comparisons needed to fix a node.

For *Ising Gauss* and $\pm J$ problems, 100 different instances were generated (for each type) using the Spin Glass Ground State server[2].

Since the aim of our experiment is to compare the influence of BP, we have not tuned the parameters to obtain the best performance in each case. We have not applied any technique such as niching or local optimization algorithms, which have been proved to be really helpful in some of these problems. As said before, our main

[2] http://www.informatik.uni-koeln.de/ls_juenger/index.html

goal is to present an exploratory work in the scope of EDAs, studying the use of BP in the sampling step.

For *OneMax*, *CheckerBoard*, and *Trap5*, 50 independent runs were executed. For the two versions of *Ising*, 10 independent runs were executed for each instance. In order to make a fair comparison, we decided to control the stochastic behavior of EDAs setting an initial seed for each run. In this way, we guarantee that EBNA and EBNA-BP start from the same point in the space (same population), and therefore the differences detected among them are due to the use of BP.

9.4.2 Results

Analyzing the results obtained from the set of experiments, no significant differences were observed in general when looking at the fitness function values. However, when analyzing the behavior of both approaches along the generations, it can be seen that BP clearly helps EBNA to obtain better individuals in earlier generations. These results agree with the experiments presented in [9]. Figures 9.2, 9.3 and 9.4 show the mean fitness function value together with the standard deviation for the five problems introduced previously.

As stated before, from a given generation on, fitness values are similar between both versions. However, for *OneMax* and *Trap5 Complete structure* EBNA-BP reaches the optimum faster (it requires around 5 generations less on average). This

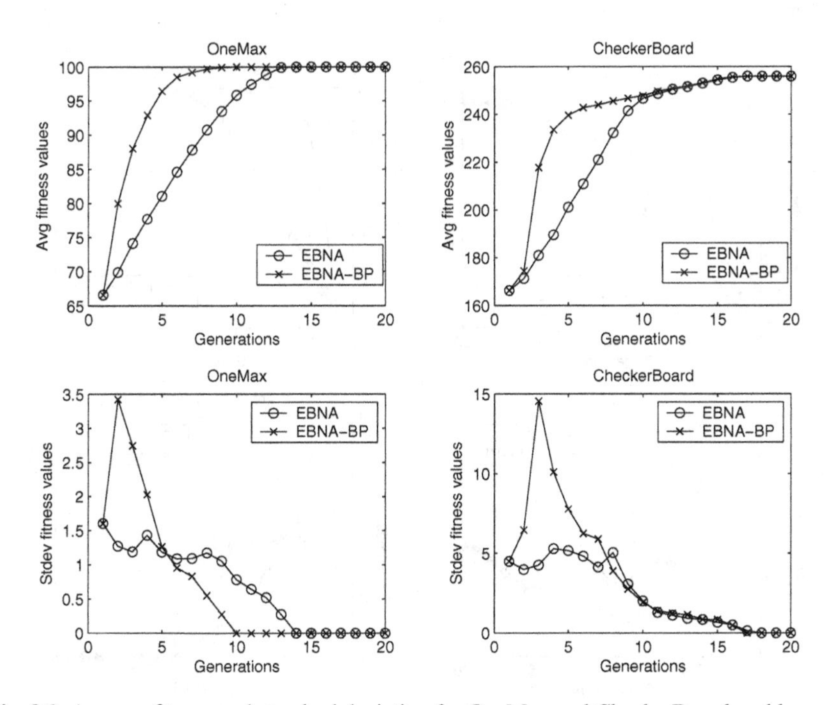

Fig. 9.2 Average fitness and standard deviation for OneMax and CheckerBoard problems.

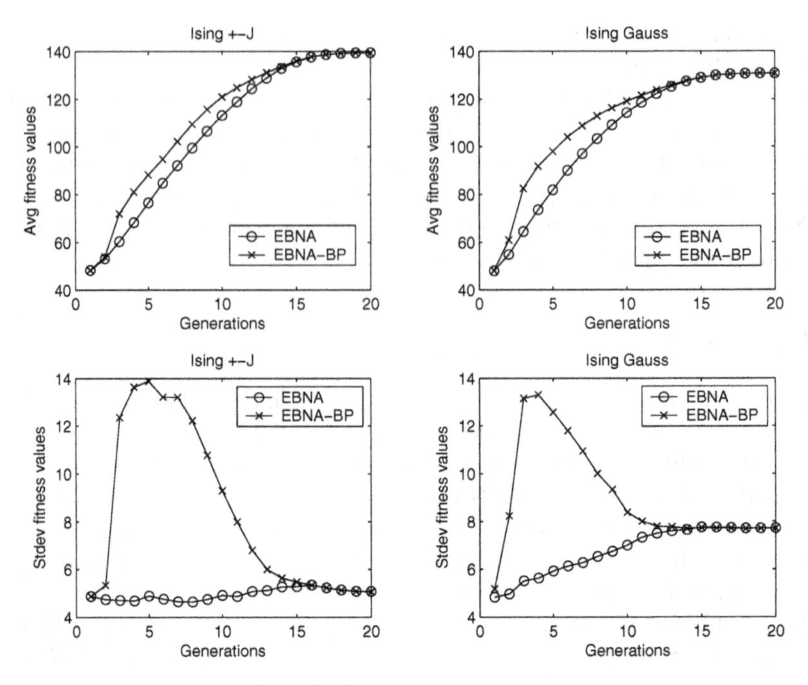

Fig. 9.3 Average fitness and standard deviation for Gaussian and $\pm J$ Ising spin glass problems.

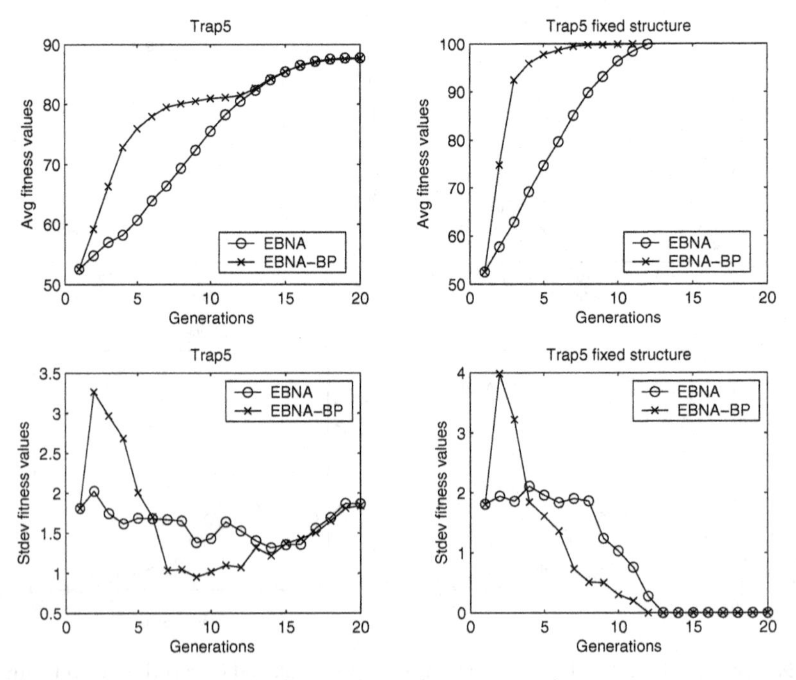

Fig. 9.4 Average fitness and standard deviation for Trap5 problem.

fact can be related to the quality of the Bayesian network learnt by EBNA. It must be noted that BP will be applied to this probabilistic model and therefore it is supposed that the better the model, the fitter the BP solution.

Another interesting aspect is the deviation of the mean best fitness values. This deviation is higher for the EBNA-BP approach. This can be related to the complexity of the Bayesian network learnt. When the factor graph has cycles, BP is not guaranteed to obtain the most probable solution. Due to this, it can happen that individuals with low probability (but with the highest fitness) will be included, changing the dynamics of the EBNA model.

9.5 Conclusions and Future Work

In this paper we have introduced a modification of the sampling procedure in EDAs. Our approach is based on factor graphs and can be used for EDAs based on Bayesian networks and other types of graphical models. Our proposal is based on the application of BP, a technique extensively used for inference in many domains. Looking at the experiments, we can conclude that BP can help EDAs to obtain better results more quickly, which is really interesting in those experiments that (1) have constraints about the number of evaluations or (2) whose evaluation function is very expensive (in terms of execution time).

In addition, we think that more research could be done in order to better understand the influence of BP. Some ideas could be the following:

- Addressing the convergence issue: As previously mentioned, it could be the case that BP does not converge, oscillating between different values. In that case it is possible to obtain an individual but there is not guarantee of it being the point with the highest probability. Some strategies to select the point in situations where similar BP algorithms oscillate have been proposed in [15]. They could be adapted to our implementation.
- Solving ties: Similarly, if there are ties in the max-marginals, it is difficult to recover the highest probability point. To fix this problem, the proposal presented by [22] could be applied.
- The sensitivity of the algorithm to the parameters: we could study the way in which the different scheduling, initial values, etc, modify the behavior of BP, and in particular its ability and time to converge.
- The dynamics of EDA-BP could be analyzed, looking at the characteristics of the individual proposed by BP, and comparing it with the other individuals in the population. In addition, different criteria could be tested to decide whether the BP individual will be included or not: always, according to diversity, etc.
- Finally, instead of factor graphs, other structures could the considered, such as region graphs, which are more expressive but also more complex [51].

In summary, we expect that work on this direction, some already underway, could provide a more efficient use of BP.

On the other hand, the criterion used to include the individual proposed by BP could be changed. In our experiments, that individual was included only when its fitness was better than the existing ones, but we could decide to always include the individual or to include it based on the genotypical distance. In addition, it would be also interesting to experiment with the k points with the highest probability. These points can be calculated using algorithms proposed by [29] and [50]. In this case, the computational cost is higher. Therefore, a balance should be found between the time spent in using BP and that of learning the Bayesian network.

Additionally, recent research on EDAs [7, 11, 40] has paid attention to the use of the models learned during the search as a source of previously unknown information about the problem. In our case, an open question is to determine the possible impact of BP in the accuracy (in terms of the mapping between the problem interactions and the dependencies learned by the model) of the models learned by EBNA.

Acknowledgement. This work has been partially supported by the Saiotek and Research Groups 2007-2012 (IT-242-07) programs (Basque Government), TIN2008-06815-C02-01, TIN2010-14931 and Consolider Ingenio 2010 - CSD2007-00018 projects (Spanish Ministry of Science and Innovation) and COMBIOMED network in computational biomedicine (Carlos III Health Institute).

References

1. Baluja, S., Davies, S.: Using optimal dependency-trees for combinatorial optimization: Learning the structure of the search space. Tech. rep., Carnegie Mellon Report, CMU-CS-97-107 (1997)
2. Bayati, M., Shah, D., Sharma, M.: Maximum weight matching via max-product belief propagation. IEEE Transactions on Information Theory 54(3), 1241–1251 (2008)
3. Chickering, D.M., Geiger, D., Heckerman, D.: Learning Bayesian networks is NP-hard. Tech. Rep. MSR-TR-94-17, Microsoft Research, Redmond, WA (1994)
4. Coughlan, J.M., Ferreira, S.J.: Finding Deformable Shapes Using Loopy Belief Propagation. In: Heyden, A., Sparr, G., Nielsen, M., Johansen, P. (eds.) ECCV 2002. LNCS, vol. 2352, pp. 453–468. Springer, Heidelberg (2002)
5. Crick, C., Pfeffer, A.: Loopy belief propagation as a basis for communication in sensor networks. In: Proceedings of the 19th Annual Conference on Uncertainty in Artificial Intelligence (UAI 2003), pp. 159–166. Morgan Kaufmann Publishers (2003)
6. Deb, K., Goldberg, D.E.: Sufficient conditions for deceptive and easy binary functions. Annals of Mathematics and Artificial Intelligence 10, 385–408 (1994)
7. Echegoyen, C., Lozano, J.A., Santana, R., Larrañaga, P.: Exact Bayesian network learning in estimation of distribution algorithms. In: Proceedings of the 2007 Congress on Evolutionary Computation, CEC 2007, pp. 1051–1058. IEEE Press (2007)
8. Echegoyen, C., Mendiburu, A., Santana, R., Lozano, J.: Analyzing the probability of the optimum in EDAs based on Bayesian networks. In: Proceedings of the 2009 Congress on Evolutionary Computation (CEC 2009), pp. 1652–1659. IEEE Press, Trondheim (2009)
9. Echegoyen, C., Mendiburu, A., Santana, R., Lozano, J.A.: Towards understanding edas based on bayesian networks through a quantitative analysis. IEEE Trans. Evolutionary Computation (accepted for publication)

10. Freeman, W.T., Pasztor, E.C., Carmichael, O.T.: Learning low-level vision. International Journal of Computer Vision 40(1), 25–47 (2000)
11. Hauschild, M., Pelikan, M., Lima, C., Sastry, K.: Analyzing probabilistic models in hierarchical BOA on traps and spin glasses. In: Thierens [48], pp. 523–530 (2007c)
12. Hauschild, M., Pelikan, M., Sastry, K., Lima, C.F.: Analyzing probabilistic models in hierarchical boa. IEEE Trans. Evolutionary Computation 13(6), 1199–1217 (2009)
13. Henrion, M.: Propagating uncertainty in Bayesian networks by probabilistic logic sampling. In: Lemmer, J.F., Kanal, L.N. (eds.) Proceedings of the Second Annual Conference on Uncertainty in Artificial Intelligence, pp. 149–164. Elsevier (1988)
14. Höns, R.: Estimation of distribution algorithms and minimum relative entropy. Ph.D. thesis, University of Bonn, Bonn, Germany (2006)
15. Höns, R., Santana, R., Larrañaga, P., Lozano, J.A.: Optimization by max-propagation using Kikuchi approximations. Tech. Rep. EHU-KZAA-IK-2/07, Department of Computer Science and Artificial Intelligence, University of the Basque Country (2007), http://www.sc.ehu.es/ccwbayes/technical.htm
16. Kschischang, F.R., Frey, B.J., Loeliger, H.A.: Factor graphs and the sum-product algorithm. IEEE Transactions on Information Theory 47(2), 498–519 (2001)
17. Larrañaga, P., Etxeberria, R., Lozano, J.A., Peña, J.: Combinatorial optimization by learning and simulation of Bayesian networks. In: Proceedings of the Sixteenth Annual Conference on Uncertainty in Artificial Intelligence (UAI 2000), pp. 343–352. Morgan Kaufmann Publishers, San Francisco (2000)
18. Larrañaga, P., Lozano, J.A. (eds.): Estimation of Distribution Algorithms. A New Tool for Evolutionary Computation. Kluwer Academic Publishers, Boston (2002)
19. Lima, C.F., Pelikan, M., Lobo, F.G., Goldberg, D.E.: Loopy Substructural Local Search for the Bayesian Optimization Algorithm. In: Stützle, T., Birattari, M., Hoos, H.H. (eds.) SLS 2009. LNCS, vol. 5752, pp. 61–75. Springer, Heidelberg (2009)
20. Lima, C.F., Pelikan, M., Sastry, K., Butz, M.V., Goldberg, D.E., Lobo, F.G.: Substructural Neighborhoods for Local Search in the Bayesian Optimization Algorithm. In: Runarsson, T.P., Beyer, H.-G., Burke, E.K., Merelo-Guervós, J.J., Whitley, L.D., Yao, X. (eds.) PPSN 2006. LNCS, vol. 4193, pp. 232–241. Springer, Heidelberg (2006)
21. McEliece, R.J., MacKay, D.J.C., Cheng, J.F.: Turbo Decoding as an Instance of Pearl's "Belief Propagation" Algorithm. IEEE Journal on Selected Areas in Communications 16(2), 140–152 (1998)
22. Meltzer, T., Yanover, C., Weiss, Y.: Globally optimal solutions for energy minimization in stereo vision using reweighted belief propagation. In: ICCV, pp. 428–435. IEEE Computer Society (2005)
23. Mendiburu, A., Santana, R., Lozano, J.: Introducing belief propagation in estimation of distribution algorithms: A parallel approach. Tech. Rep. EHU-KAT-IK-11-07, Department of Computer Science and Artificial Intelligence, The University of the Basque Country (2007)
24. Mendiburu, A., Santana, R., Lozano, J.A., Bengoetxea, E.: A parallel framework for loopy belief propagation. In: Thierens [48], pp. 2843–2850 (2007c)
25. Mühlenbein, H.: The equation for response to selection and its use for prediction. Evolutionary Computation 5(3), 303–346 (1997)
26. Mühlenbein, H., Höns, R.: The factorized distributions and the minimum relative entropy principle. In: Pelikan, M., Sastry, K., Cantú-Paz, E. (eds.) Scalable Optimization via Probabilistic Modeling: From Algorithms to Applications. SCI, pp. 11–38. Springer (2006)

27. Mühlenbein, H., Mahnig, T.: FDA – a scalable evolutionary algorithm for the optimization of additively decomposed functions. Evolutionary Computation 7(4), 353–376 (1999)
28. Mühlenbein, H., Paaß, G.: From Recombination of Genes to the Estimation of Distributions I. Binary Parameters. In: Ebeling, W., Rechenberg, I., Voigt, H.-M., Schwefel, H.-P. (eds.) PPSN 1996. LNCS, vol. 1141, pp. 178–187. Springer, Heidelberg (1996)
29. Nilsson, D.: An efficient algorithm for finding the M most probable configurations in probabilistic expert systems. Statistics and Computing 2, 159–173 (1998)
30. Ochoa, A.: EBBA - Evolutionary best basis algorithm. In: Ochoa, A., Soto, M.R., Santana, R. (eds.) Proceedings of the Second International Symposium on Adaptive Systems (ISAS 1999), pp. 93–98. Editorial Academia, Havana (1999)
31. Ochoa, A., Höns, R., Soto, M., Mühlenbein, H.: A Maximum Entropy Approach to Sampling in EDA – the Single Connected Case. In: Sanfeliu, A., Ruiz-Shulcloper, J. (eds.) CIARP 2003. LNCS, vol. 2905, pp. 683–690. Springer, Heidelberg (2003)
32. Pearl, J.: Probabilistic Reasoning in Intelligent Systems. Morgan Kaufmann, Palo Alto (1988)
33. Pelikan, M., Goldberg, D.E., Cantú-Paz, E.: BOA: The Bayesian optimization algorithm. In: Banzhaf, W., Daida, J., Eiben, A.E., Garzon, M.H., Honavar, V., Jakiela, M., Smith, R.E. (eds.) Proceedings of the Genetic and Evolutionary Computation Conference, GECCO 1999, Orlando FL, vol. I, pp. 525–532. Morgan Kaufmann Publishers, San Francisco (1999)
34. Pelikan, M., Goldberg, D.E., Lobo, F.: A survey of optimization by building and using probabilistic models. IlliGAL Report No. 99018, University of Illinois at Urbana-Champaign, Illinois Genetic Algorithms Laboratory, Urbana, IL (1999)
35. Pelikan, M., Sastry, K., Goldberg, D.E.: Sporadic model building for efficiency enhancement of the hierarchical BOA. Genetic Programming and Evolvable Machines 9(1), 53–84 (2008)
36. Pereira, F.B., Machado, P., Costa, E., Cardoso, A., Ochoa, A., Santana, R., Soto, M.R.: Too busy to learn. In: Proceedings of the 2000 Congress on Evolutionary Computation, CEC 2000, pp. 720–727. IEEE Press, La Jolla Marriott Hotel La Jolla (2000)
37. Potetz, B.: Efficient belief propagation for vision using linear constraint nodes. In: 2007 IEEE Conference on Computer Vision and Pattern Recognition, pp. 1–8. IEEE (2007)
38. Richardson, T.J., Urbanke, R.L.: The capacity of low-density parity-check codes under message-passing decoding. IEEE Transactions on Information Theory 47(2), 599–618 (2001)
39. Santana, R.: Advances in probabilistic graphical models for optimization and learning: Applications in protein modelling. Ph.D. thesis (2006)
40. Santana, R., Larrañaga, P., Lozano, J.A.: The Role of a Priori Information in the Minimization of Contact Potentials by Means of Estimation of Distribution Algorithms. In: Marchiori, E., Moore, J.H., Rajapakse, J.C. (eds.) EvoBIO 2007. LNCS, vol. 4447, pp. 247–257. Springer, Heidelberg (2007)
41. Santana, R., Larrañaga, P., Lozano, J.A.: Protein folding in simplified models with estimation of distribution algorithms. IEEE Transactions on Evolutionary Computation (2008) (in Press)
42. Santana, R., Larrañaga, P., Lozano, J.A.: Learning factorizations in estimation of distribution algorithms using affinity propagation. Evolutionary Computation 18(4), 515–546 (2010)

43. Sastry, K., Goldberg, D.E.: Designing Competent Mutation Operators Via Probabilistic Model Building of Neighborhoods. In: Deb, K., et al. (eds.) GECCO 2004. LNCS, vol. 3103, pp. 114–125. Springer, Heidelberg (2004)
44. Sastry, K., Lima, C., Goldberg, D.E.: Evaluation relaxation using substructural information and linear estimation. In: Proceedings of the 8th annual Conference on Genetic and Evolutionary Computation, GECCO 2006, pp. 419–426. ACM Press, New York (2006)
45. Sastry, K., Pelikan, M., Goldberg, D.: Efficiency enhancement of genetic algorithms via building-block-wise fitness estimation. In: Proceedings of the 2004 Congress on Evolutionary Computation, CEC 2004, pp. 720–727. IEEE Press, Portland (2004)
46. Shakya, S.: DEUM: A framework for an estimation of distribution algorithm based on markov random fields. Ph.D. thesis, The Robert Gordon University, Aberdeen, UK (2006)
47. Soto, M.R.: A single connected factorized distribution algorithm and its cost of evaluation. Ph.D. thesis, University of Havana, Havana, Cuba (2003) (in Spanish)
48. Thierens, D. (ed.): Proceedings of Genetic and Evolutionary Computation Conference, GECCO 2007, London, England, UK, July 7-11. Companion Material. ACM (2007)
49. Wainwright, M., Jaakkola, T., Willsky, A.: Tree consistency and bounds on the performance of the max-product algorithm and its generalizations. Statistics and Computing 14, 143–166 (2004)
50. Yanover, C., Weiss, Y.: Finding the M most probable configurations using loopy belief propagation. In: Thrun, S., Saul, L., Schölkopf, B. (eds.) Advances in Neural Information Processing Systems, vol. 16, p. 289. MIT Press, Cambridge (2004)
51. Yedidia, J.S., Freeman, W.T., Weiss, Y.: Constructing free energy approximations and generalized belief propagation algorithms. IEEE Transactions on Information Theory 51(7), 2282–2312 (2005)

Chapter 10
Continuous Estimation of Distribution Algorithms Based on Factorized Gaussian Markov Networks

Hossein Karshenas, Roberto Santana, Concha Bielza, and Pedro Larrañaga

Abstract. Because of their intrinsic properties, the majority of the estimation of distribution algorithms proposed for continuous optimization problems are based on the Gaussian distribution assumption for the variables. This paper looks over the relation between the general multivariate Gaussian distribution and the popular undirected graphical model of Markov networks and discusses how they can be employed in estimation of distribution algorithms for continuous optimization. A number of learning and sampling techniques for these models, including the promising regularized model learning, are also reviewed and their application for function optimization in the context of estimation of distribution algorithms is studied.

10.1 Introduction

Approaches to continuous optimization with estimation of distribution algorithms (EDAs) [24], can be divided into two general categories: (i) Discretization of problem domain and then application of discrete EDAs [49]; (ii) Direct application of EDAs based on continuous probabilistic models [3, 5, 13].

In the latter approach, Gaussian distributions have been the probabilistic model of choice in most of the research in this area, considering either non-overlapping factorized distributions (e.g., Gaussian UMDA) or distributions defined by dependencies encoded in graphical models (e.g., Gaussian networks). The research on the use of undirected graphical models (Markov networks) in EDAs however, has been mainly focused on discrete domain optimization [39, 43–45]. In this paper,

Hossein Karshenas · Concha Bielza · Pedro Larrañaga
Computational Intelligence Group, Faculty of Informatics, Technical University of Madrid
e-mail: {hkarshenas,mcbielza,pedro.larranaga}@fi.upm.es

Roberto Santana
Intelligent Systems Group, Faculty of Informatics,
University of the Basque Country (UPV/EHU), San-Sebastian, Spain
e-mail: roberto.santana@ehu.es

S. Shakya and R. Santana (Eds.): Markov Networks in Evolutionary Computation, ALO 14, pp. 157–173.
springerlink.com

continuous EDAs based on Gaussian distribution are analyzed from the Gaussian Markov random field [46] perspective, a probabilistic graphical model successfully applied for handling uncertainty in many practical domains. It is shown that the analysis of undirected graphical models, as it is done in discrete Markov network-based EDAs, can be also extended to continuous domains. An important role in this analysis is played by the precision matrix which connects the dependencies between the variables to the Gaussian distributions. We focus on marginal product factorizations as a particular case of undirected graphical models. Next section will discuss the theoretical issues related to Gaussian distributions and their relation to Gaussian Markov random fields. Section 10.3 describes the main steps of the introduced algorithm. Special attention is given to different variants of incorporating regularization into model learning and the description of affinity propagation, the clustering technique used by the proposed EDA for finding the factorization of the distribution. Section 10.4 discusses previous works related to our proposal. The experiments that illustrate the behavior of the introduced approach using two different types of functions are presented in Section 10.5. Finally, Section 10.6 concludes the paper.

10.2 Multivariate Gaussian Distribution

Multivariate Gaussian distribution (MGD) is the most frequently used probability distributions for continuous optimization problems in estimation of distribution algorithms [2, 4, 28]. A multivariate Gaussian distribution $\mathcal{N}(\mu,\Sigma)$ over a vector of n random variables $X = (X_1,\ldots,X_n)$ is defined with two parameters: μ is the n-dimensional vector of mean values for each variable, and Σ is a positive-definite and symmetric $n \times n$ covariance matrix.

A square covariance matrix Σ is positive definite if $x\Sigma x^T > 0$, $\forall x \in \mathbb{R}^n \setminus \{0\}$. Positive definite matrices are guaranteed to be full-ranked and non-singular. When the '$>$' relation is replaced with '$\geq$', the square matrix becomes positive semi-definite which can be singular. Although in general MGDs can have positive semi-definite covariance matrices, but since only positive definite covariance matrices are invertible, this type of MGDs are only considered in this paper.

Geometrically, MGDs specify a set of parallel ellipsoidal contours around the mean vector. The mean vector determines the bias of each variable's values from origin and the variances, i.e., entries along the diagonal of the covariance matrix, are responsible for specifying the spread of values. The covariances (i.e., the off-diagonal entries in the covariance matrix) determine the shape of the ellipsoids. The mean vector and the covariance matrix are computed as the first two moments of the Gaussian distribution: (considering row-wise vectors)

$$\mu = E(X)$$
$$\Sigma = E((X-\mu)^T(X-\mu)) = E(X^T X) - \mu^T \mu$$

The typical representation of an MGD, sometimes referred to as the *moment* form, is given by

$$p_{\mathscr{N}(\mu,\Sigma)}(x) = \frac{1}{\sqrt{(2\pi)^n|\Sigma|}} \exp\left(-\frac{1}{2}(x-\mu)\Sigma^{-1}(x-\mu)^T\right) \tag{10.1}$$

This equation can be transformed to the *information* form representation of MGD [25], also known as the canonical or natural form

$$p_{\mathscr{N}^{-1}(h,\Theta)}(x) = \frac{\exp\left(-\frac{1}{2}h\Theta^{-1}h^T\right)}{\sqrt{(2\pi)^n|\Theta^{-1}|}} \exp\left(-\frac{1}{2}x\Theta x^T + xh^T\right) \tag{10.2}$$

where $h = \mu\Sigma^{-1}$ is called the potential vector and $\Theta = \Sigma^{-1}$ is the inverse covariance matrix, known as the *precision*, concentration or information matrix. For a valid MGD represented in the information form, the precision matrix should be positive definite.

10.2.1 Markov Networks

Similar to other probabilistic graphical models, a Markov network $\mathscr{M}(\mathscr{G},\Phi)$ is composed of two components: (i) Graphical structure $\mathscr{G}$ and (ii) Parameters Φ. The nodes in the structure represent the variables, and the undirected edges correspond to probabilistic interactions between the neighboring nodes.

The parameters of a Markov network represent the affinities between related variables and the compatibility of their values. They are represented with factors $\phi_k \in \Phi$ (non-negative functions) that are defined over network cliques (complete subgraphs) $C_k \subseteq X$. The normalized product of these factors (according to factors multiplication rule) define the so called *Gibbs* distribution, factorized over the Markov network, which gives the joint probability distribution encoded in the network

$$p_{\Phi}(x) = \frac{1}{Z} \prod_{\phi_k \in \Phi, C_k \in \mathscr{C}} \phi_k(x_{C_k}) \tag{10.3}$$

where Z is the normalization term usually called the *partition* function, and $\mathscr{C}$ is a subset of network cliques that covers all variables in X.

10.2.2 Gaussian Markov Random Fields

When the interactions between the variables are symmetrical and there is no specific direction for the influence of the variables over each other, an undirected graph is more appropriate for representing the correlations. Markov networks are a type of probabilistic graphical models fitted to this need.

A widely used class of Markov networks is the pairwise Markov networks [15], where all of the factors are defined over either single or pairs of variables. More specifically, a pairwise Markov network has two types of factors:

i) Node factors $\phi_i(X_i)$ defined over every single variable X_i,
ii) Edge factors $\phi_{ij}(X_i,X_j)$ defined over the ending variables X_i and X_j of every edge.

MGDs and pairwise Markov networks are closely related. The precision matrix of an MGD represents partial covariances between pairs of variables variables. A zero value in any entry θ_{ij} of this matrix implies that the corresponding two variables are conditionally independent given all other variables and vice versa:

$$\theta_{ij} = 0 \quad \Longleftrightarrow \quad (X_i \perp X_j \mid X \setminus \{X_i,X_j\}) \in \mathscr{I}_p$$

where $\mathscr{I}_p$ represents the set of conditional independencies satisfied by MGD. This type of conditional independence is exactly the pairwise Markov property [29] encoded by Markov networks, and therefore the zero pattern of the precision matrix directly induces a Markov network, which is known as *Gaussian Markov Random Field* (GMRF) [38].

The structure of this Markov network is obtained by introducing an edge between every two nodes whose corresponding variables are partially correlated (a non-zero entry in the precision matrix). The parameters of the model are node and edge factors obtained, for example, by decomposing the variable part of the exponential factor of the MGD in Equation (10.2) to terms consisting of single and pairs of variables

$$\begin{aligned}\exp\Big(-\frac{1}{2}X\Theta X^T + Xh^T\Big) = &\Big[\exp(-\frac{1}{2}\theta_{11}X_1^2 + h_1X_1)\cdots \\ &\exp(-\frac{1}{2}\theta_{nn}X_n^2 + h_nX_n)\Big]\cdot\Big[\exp(-\theta_{12}X_1X_2)\cdots \\ &\exp(-\theta_{1n}X_1X_n)\cdots\exp(-\theta_{n-1,n}X_{n-1}X_n)\Big]\end{aligned}$$

The joint Gibbs distribution factorizing over this Markov network (Equation (10.3)) can then be obtained by computing the partition function

$$Z = \int_{-\infty}^{\infty} \exp\Big(-\frac{1}{2}X\Theta X^T + Xh^T\Big)\,dX = \frac{\sqrt{(2\pi)^n|\Theta^{-1}|}}{\exp\big(-\frac{1}{2}h\Theta^{-1}h^T\big)}.$$

While transforming an MGD to its corresponding GMRF is straightforward, thanks to the information representation in Equation (10.2), it should be noted that it is not possible to obtain a valid MGD from every pairwise Markov network with log-quadratic factors. This is because not every pairwise Markov network with log-quadratic factors will result in a positive definite precision matrix [25].

10.2.3 Marginal Product Factorizations

Factorization of the joint probability distribution can represent both marginal and conditional independence relationships between the variables. If the factorization is only based on the marginal independence relationships between the sets of variables then it is called a marginal product factorization. Marginal product factorizations can be represented with directed and undirected probabilistic graphical models. In particular, when a GMRF represents a marginal product model (MPM), its cliques do not overlap and therefore it is possible to independently estimate the probability distribution for the variables in each clique.

In this paper, we consider EDAs using MPM representation of the probability distribution with GMRFs. This is appropriate when variables in the problem can be divided into a number of disjoint groups. Many of the real-world problems actually consist of several smaller components that are either independent or weakly connected. This type of GMRFs also allows to evaluate the learning algorithms we introduced in this chapter. In the discrete case, two well known examples of EDAs that use MPMs are the univariate marginal distribution algorithm (UMDA) [34] and the extended compact genetic algorithm (ECGA) [16]. In Section 10.4 we also review some of the continuous EDAs based on marginal product factorizations.

10.3 GMRF-Based EDA with Marginal Factorization

We have seen that GMRFs can represent independence relationships between continuous variables. The abstraction of the problem regularities captured by these undirected graphical models can be useful for continuous optimization in EDAs. A necessary step in incorporating undirected models in optimization is designing algorithms to feasibly learn models from data, from memory and time constraints points of view.

In this section we propose an approach for learning a subclass of GMRFs that represent MPMs. There are several alternatives for learning these undirected continuous models within EDAs, from which some are:

i) Estimation of MGD's covariance or precision matrix.
ii) Learning the structure of GMRF and its factors.
iii) Hybrid approaches.

The first approach includes maximum likelihood estimation as well as other covariance matrix selection techniques discussed in the literature [37, 50, 51]. The methods for inducing an undirected (in)dependence structure between variables are the typical choice in the second approach. Usually these methods use techniques like statistical hypothesis tests or mutual information (entropy) between variables to decide about their (in)dependence. Based on these (in)dependencies, the local neighborhood of each variable is obtained which can then be combined to obtain a full structure and compute the factors for the related variables. The third class of methods combines the computation of the covariance or precision matrix with the identification of the neighborhood structure of the variables. The approach introduced in this chapter belongs to the third class of methods.

10.3.1 A Hybrid Approach to Model Learning of GMRF

The main idea of our algorithm is to initially identify the putative neighborhood of each variable X_i by learning a regularized regression model [19, 48]. In this model, the dependence of X_i on each of the variables in $X \setminus X_i$ is represented with a weight. In the second step, variables are clustered into disjoint factors according to the strength of their dependence weights. Finally, an MGD is estimated for each factor using one of the methods in the first approach. The main steps of the proposed method are presented in Algorithm 10.1. Each of these steps are explained in detail in the following sections.

---

1. Set $t \Leftarrow 0$. Generate M points randomly.
2. **for** each variable
3. Compute its linear dependence weights on all other variables
4. Cluster the variables into disjoint cliques using the weight matrix
5. **for** each clique
6. Estimate an MGD for the variables in the clique

Fig. 10.1 Hybrid GMRF-EDA learning algorithm

10.3.2 Model Learning Using Regularization

Regularized model learning is one of the promising methods proposed in statistics for estimating a more accurate and sparser model. In regularization, the value of model parameters are penalized to shrink them toward zero. This is done by adding a regularization term of the form $\lambda J(\beta)$ to model estimation, where λ is the regularization parameter and $J(\cdot)$ is a function of model parameters (β_i). Some of the regularization techniques have the interesting property of setting some of the model parameters exactly equal to zero, thus implicitly performing variable selection [48].

The hybrid GMRF learning method proposed in this chapter, uses regularized linear regression models to find the dependencies between the variables. The parameters of the model learnt for each variable X_i, are the weights specifying the influence of other variables on the values of X_i. Thanks to regularized estimation, some of these weights will be exactly zero, suggesting that the two variables are independent, given the rest of variables. The regularization techniques considered in this study are:

- LASSO (lasso) or ℓ_1 regularization [48]: This technique penalizes the absolute values of the model parameters (regression coefficients), and can act as a variable selection operator. It is one of the most frequently used regularization techniques and different variants have been proposed for it in the literature.

- Elastic net (elnet) [52]: The penalization term in this technique is a linear combination of the previous ℓ_1 regularization term with an ℓ_2 regularization term [20], causing a grouping effect where correlated variables will be in or out of the model together.
- Least angle regression (lars) [9]: In this method the variables are added to the model (setting their coefficients to non-zero) one at a time based on their correlation with the error of estimation, and in exactly N steps, where N is the size of the learning data. This algorithm is able to compute the whole solution path (all possible coefficients configurations) of a regression model for different λ values very efficiently.

Regularization can also be applied for obtaining a regularized estimation of the covariance matrix or its inverse. In these methods, sparser estimation of covariance or precision matrices are obtained by applying regularization on their entries, causing marginal or conditional independencies between variables. To evaluate the proposed hybrid model learning algorithm, it is also compared with the following two regularization methods used for obtaining a regularized estimation of the covariance matrix:

- Shrinkage estimation (Shrink) [42]: In this method the sample covariance matrix, obtained from maximum likelihood estimation, is shrunk towards a target covariance matrix with smaller number of free parameters (e.g., a diagonal matrix). This method uses analytical relations to compute the proper value of the regularization parameter and can result in a statistically more efficient estimation of the covariance matrix.
- Graphical LASSO (Glasso) algorithm [12]: This method is based on obtaining a regularized maximum likelihood estimation of the MGD. An ℓ_1 regularization is applied on the entries of the precision matrix, forcing sparser Markov network structures. Here, the regularization parameter, determining the strength of penalization is left as an open parameter.

10.3.3 Clustering by Affinity Propagation

Clustering methods are used to group objects into different sets or clusters, in such a way that each cluster comprises similar objects. Clusters can then be associated to labels that are used to describe the data and identify their general characteristics. Among the the best known clustering algorithms are k-means [17] and k-center clustering [1].

Affinity propagation (AP) [11] is a recently introduced clustering method which takes as input a set of measures of similarity between pairs of data points and outputs a set of clusters of those points. For each cluster, a typical or representative member, which is called exemplar, is identified. We use affinity propagation as an efficient way to find the Markov network neighborhood of the variables from the linear weights output by regularized regression methods.

AP takes as input a matrix of similarity measures between each pair of points and a set of preferences which are a measure of how likely each point is to be chosen as exemplar. The algorithm works by exchanging messages between the points until a stop condition, which reflects an agreement between all the points with respect to the current assignment of the exemplars, is satisfied. These messages can be seen as the way the points share local information in the gradual determination of the exemplars. For more details on AP, see [11]. In the context of GMRF-EDA, each variable will correspond to a point and the similarity between two points X_i and X_j will be the absolute value of weight w_{ij} obtained from the regularized regression model of variable X_i. Since in general $w_{ij} \neq w_{ji}$, the similarity matrix is not symmetric. AP also takes advantage of the sparsity of the similarity matrix obtained from regularized estimation, when such distribution of similarity values is available. The message-passing procedure may be terminated after a fixed number of iterations, when changes in the messages fall below a threshold, or after the local decisions stay constant for some number of iterations.

10.3.4 Estimating an MGD for the Factors

In the final step of the proposed hybrid model learning algorithm, an MGD of the variables in each factor is estimated. We expect that the factorized distribution obtained by this model estimation will give a more accurate estimation of the target MGD (of all variables) in comparison to learning a single multivariate distribution for all of the problem variables.

10.3.5 Sampling Undirected Graphical Models for Continuous Problems

Sampling is one of the most problematic steps in EDAs based on undirected graphical models. The problem is mainly related to the loops existing in the model structure that do not allow a straightforward application of simple sampling techniques like the probabilistic logic sampling (PLS) method [18]. The application of Markov Chain Monte Carlo (MCMC) methods like Gibbs sampling [14] also has a high computational complexity.

Obtaining decomposable approximations of the Markov networks to allow the application of PLS algorithm [39], merging cliques of the Markov network to capture as many dependencies as possible before applying exact or approximate sampling algorithms [21], and computation of the most probable configurations based on belief propagation [33] are among other options for sampling undirected graphical models that have been used in discrete EDAs and could be applied to continuous problems. Here, we independently sample the MGDs for each of the factors.

10.4 Related Work

A number of works have proposed the use of MPMs for continuous problems. In [10] and [30] two different algorithms are proposed that learn variants of ECGA for real-valued problems. Both algorithms are based on discretizing the continuous variables previous to the construction of the MPM. Lanzi et al. [26] propose an ECGA that instead of mapping real values into discrete symbols, models each cluster of variables using a multivariate probability distribution and guides the partitioning process using an MDL metric for continuous distributions. To learn the clusters of variables, an adaptive clustering algorithm, namely the BEND random leader algorithm, is used.

Recently, Dong et al. [8] have proposed to compute the correlation matrix as an initial step to do a coarse learning such as identifying weakly dependent variables. These variables are independently modeled using a univariate Gaussian distribution while the other variables are randomly projected into subspaces (clusters) that are modeled using a multivariate Gaussian distribution.

Although the introduction of regularization techniques in EDAs is very recent, different variants of its application have been tried. In particular, the approach in which the local neighborhood of each variable is modeled with a regularized regression method and then a specific combination strategy is applied to aggregate these models [32] has been applied in different contexts. It has been incorporated into EDAs for optimization based on undirected graphical models both for discrete [40] and continuous domains [23]. In [31], the task of selecting the proper structure of the Markov network is addressed by using l1-regularized logistic regression techniques. In [35], the class of shrinkage estimation of distribution algorithms has been introduced. The results presented there show that shrinkage regularization can dramatically reduce the critical population size needed by EMNA in the optimization of continuous functions.

AP has been previously applied in evolutionary computation. It has been used to learn MPMs in the learning phase of EDAs that learn discrete MPMs [41]. In a different context, it has been applied as a niching procedure for EDAs based on Markov chain models [6].

10.5 Experiments

The main objective of our experiments is to study the proposed GMRF-EDAs in learning a factorized distribution model and compare their behavior to that of EDAs that use regularization methods for learning a single Gaussian distribution. The results are also compared with those of UMDAc [27], the EDA that assumes total independence between the variables. We also analyze the accuracy of the learning methods in recovering an accurate structure of the problem.

10.5.1 Benchmark Functions

To evaluate the performance of the algorithms, we use two classes of functions that represent completely different domains of difficulty in terms of optimization: An additive deceptive function and a simplified protein folding model.

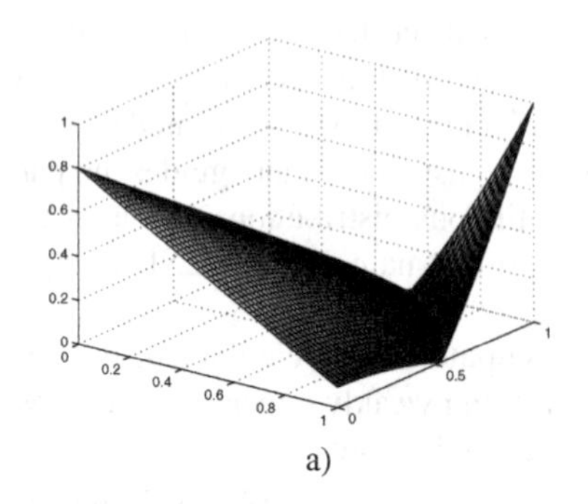

a)

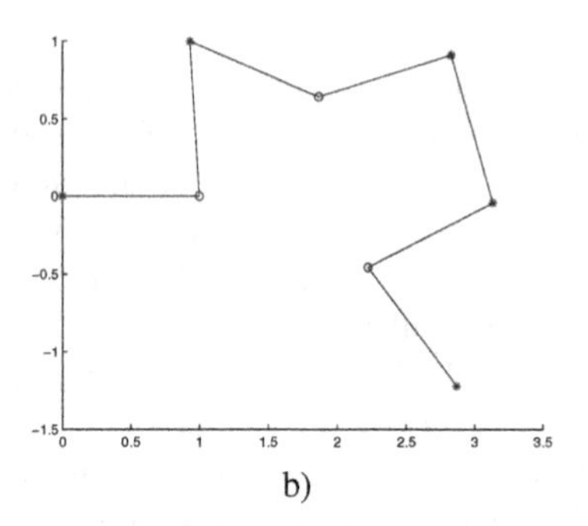

b)

Fig. 10.2 a) 2-dimensional continuous trap function. b) One possible configuration of the Fibonacci sequence $S_5 = BABABBAB$.

The 2D-deceptive function [36] is composed of the aggregation of 2-dimensional trap functions which have a local optimum with a large basin of attraction and an isolated global optimum (Fig. 10.2a)), and should be maximized.

$$f_{2D-deceptive}(\mathbf{x}) = \sum_{i=1}^{n/2} f_{2D-trap}(x_{2i-1}, x_{2i})$$

where

$$f_{2D-trap}(x,y) = \begin{cases} 0.8 - \sqrt{\frac{x^2+y^2}{2}} & \text{if } \sqrt{\frac{x^2+y^2}{2}} \leq 0.8 \\ -4 + 5\sqrt{\frac{x^2+y^2}{2}} & \text{otherwise} \end{cases}.$$

Off-lattice models are simplified protein models that, in opposition to the HP simplified model [7], do not follow a given lattice topology. Instead, the 2D or 3D coordinate in the real axis define the positions of the protein residues. Among the off-lattice models with known lowest energy states is the *AB* model [47], where *A* stands for hydrophobic and *B* for polar residues. The distances between consecutive residues along the chain are held fixed to $b = 1$, while non-consecutive residues interact through a modified Lennard-Jones potential. There is an additional energy contribution from each angle θ_i between successive bonds. The energy function for a chain of n residues, that is to be minimized, is shown in equation (10.4).

$$E = \sum_{i=2}^{n-1} E_1(\theta_i) + \sum_{i=1}^{n-2} \sum_{j=i+2}^{n} E_2(r_{ij}, \zeta_i, \zeta_j), \tag{10.4}$$

where

$$E_1(\theta_i) = \frac{1}{4}(1 - cos\theta_i) \tag{10.5}$$

and

$$E_2(r_{ij}, \zeta_i, \zeta_j) = 4(r_{ij}^{-12} - C(\zeta_i, \zeta_j) r_{ij}^{-6}) \tag{10.6}$$

Here, r_{ij} is the distance between residues i and j (with $i < j$). Each ζ_i is either A or B, and $C(\zeta_i, \zeta_j)$ is $+1$, $+\frac{1}{2}$, and $-\frac{1}{2}$ respectively, for *AA*, *BB*, and *AB* pairs, giving strong attraction between *AA* pairs, weak attraction between *BB* pairs, and weak repulsion between *A* and *B* [22].

We only consider Fibonacci sequences defined recursively by

$$S_0 = A, \quad S_1 = B, \quad S_{i+1} = S_{i-1} * S_i \tag{10.7}$$

where $*$ is the concatenation operator. Fig. 10.2b) shows a possible configuration for sequence $S_5 = BABABBAB$.

A 2D off-lattice solution of the *AB* model can be represented as a set of $n-2$ angles. Angles are represented as continuous values in $[0, 2\pi]$. The first two residues can be fixed at positions $(0,0)$ and $(1,0)$. The other $n-2$ residues are located from the angles with respect to the previous bond. We look for the set of angles that defines the optimal off-lattice configuration. As instances, we consider Fibbonacci sequences for numbers $(5,6,7,8)$. The respective sizes of these sequences, in the same order, are $n \in (13,21,34,55)$.

10.5.2 EDA Parameters

All of the EDA variants used truncation selection with $\tau = 0.5$. The population size was $5n$ and the maximum number of allowed generations was 2000. We conduct 30 experiments for small instances ($n \leq 30$) and 15 experiments for larger instances.

10.5.3 Influence of the Regularization Parameter

In the first experiment, we investigated the accuracy of the structural learning algorithm as a function of the regularization parameter (λ). Analyzing the influence of the regularization parameter is very important to understand the way in which the introduction of regularization affects the behavior of the algorithms. We did this analysis for the 2D-deceptive function for which the problem structure is known.

Fig. 10.3 represents the typical precision matrices obtained at the end of an EDA run using regularized covariance matrix estimation, with two different configuration of this parameter, on a 10 variable 2D-deceptive function with cross-related variables (i.e. first with last, second with penultimate, etc.).

Fig. 10.3a) uses a constant penalization value throughout the whole EDA run while the other (Fig. 10.3b)) starts with a small value and dynamically increases it along the evolution. As it can be seen, the employed regularization technique allows the algorithm to obtain a very good estimation of the precision matrix and its corresponding GMRF structure.

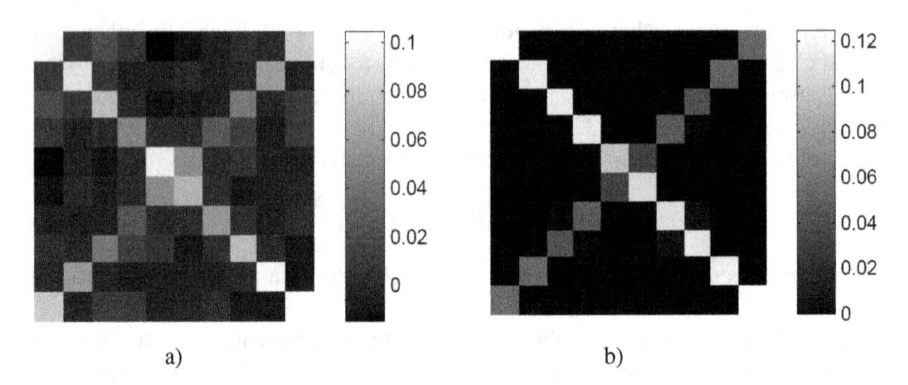

Fig. 10.3 Precision matrices learned by two different learning methods. a) $\lambda = 0.001$. b) Dynamic λ.

10.5.4 Optimization Performance

In the second step we compare the results of different GMRF-EDAs for the problems selected. Fig. 10.4 shows the average best value of the 2D-deceptive function ($n = 30$) in each generation. The average has been computed from the set of all experiments conducted. It can be appreciated that UMDAc, which considers a fixed total factorization of the distribution, starts outperforming other algorithms in the first generations but as the evolution continues, the initial lead of this algorithm is lost. On the other hand, EDAs based on regularized model learning using the graphical LASSO and shrinkage estimation methods, which consider no explicit factorization of the distribution, have a poor behavior and are not able to reach the best solutions achieved by UMDAc. The GMRF-EDAs using the proposed hybrid model learning algorithm, are able to constantly improve the average best value and obtain better results and the end of optimization.

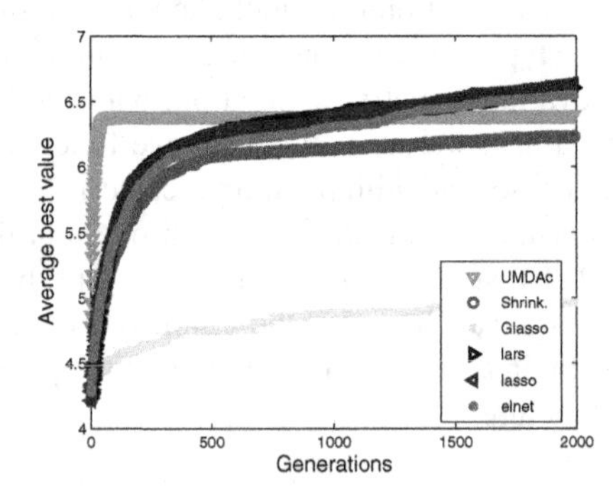

Fig. 10.4 Results of different EDAs for the $f_{2D-deceptive}$ function.

We also investigate the behavior of the algorithms for the off-lattice protein folding model. It is worth to mention that this is a very hard problem where the structural interactions between the variables are not clearly defined. There seems to be a clear dependence between adjacent variables in the AB sequence. However, the way in which other interactions between the AB residues are translated to dependencies in the model is not clear.

Fig. 10.5 shows the average best value of the off-lattice protein folding models corresponding to Fibonacci sequences 6 and 7 in each generation. For the first problem shown in (Fig. 10.5a)), it can be seen that UMDAc is outperformed by all other EDAs from the initial stages of the search. In this instance, the best results are achieved when estimating a single MGD with graphical LASSO, although the MPM learnt with elastic net regularization technique closely follows.

On the second instance of off-lattice protein problem (Fig. 10.5b)), the performance of UMDAc gets very close to EDAs using graphical LASSO and shrinkage estimation methods, and outperforming GMRF-EDAs that try to decompose the problem by learning dependencies between variables. There seems to be an important variability between the characteristics of the different off-lattice protein instances. In some situations such as the Fibonacci sequence number 6, capturing dependencies between the variables of the problem contributes to improve the quality of the obtained solutions. However, there are cases where explicitly learning the dependencies does not improve the results of the simpler univariate models.

Another interesting issue is to observe the disparate behavior exhibited by some EDAs like the one that uses graphical LASSO. It behaves very different in comparison to all other algorithms for 2D-deceptive function, achieving the worst results. Nevertheless, it reaches the best results for the off-lattice protein models. More research on this type of regularized model learning is needed to discern which mechanisms explain the difference between the performance of the algorithm for these two classes of functions.

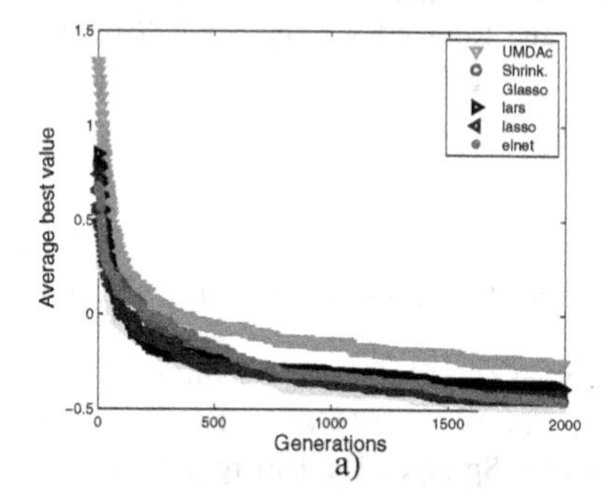

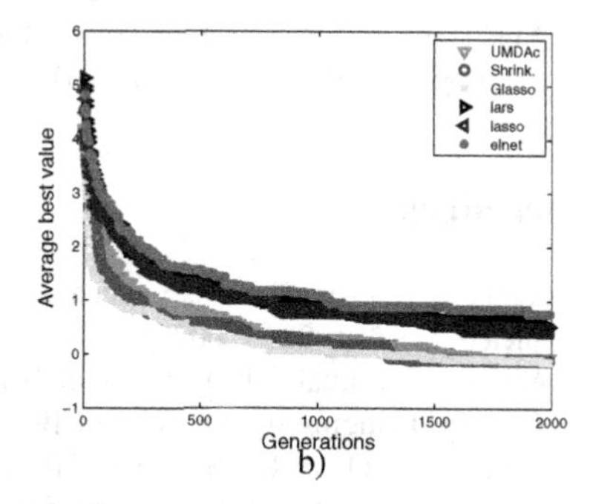

Fig. 10.5 Results of different EDAs for two different instances of the Off-lattice AB models. a) Sequence s_6. b) Sequence s_7.

10.6 Conclusions

This paper discussed how the multivariate Gaussian distribution, used by many continuous estimation of distribution algorithms, has a close relationship with a special type of Markov networks, namely the Gaussian Markov random field. It was shown how the (in)dependencies encoded in the precision matrix of the Gaussian distribution corresponds to the edges in the structure of the induced Markov network, and how factors of the network can be driven from the joint probability density function.

Based on this correspondence, some of the topics in learning and sampling these probabilistic graphical models, when employed in EDAs, were discussed. Specifically, some of the methods that can be used to approach the learning and sampling of Gaussian Markov random fields were presented. Regularized model learning was pointed out as a promising method for this purpose, especially in high dimensional problems.

We have introduced a GMRF-EDA that combines regularized regression with affinity propagation to learn regularized marginal product models. This is a different approach to learn more accurate MPMs for continuous problems. We have presented preliminary results on two different functions but more extensive experimentation is needed to determine how the characteristics of the regularization method influence the outcome of the EDAs.

It is worth mentioning that EDAs that learn undirected graphical models do not necessarily need to stick to Gaussian distributions and can employ other types of multivariate probability distributions once they have obtained the structure. The proposed hybrid model learning algorithm can be straitforwardly adopted for this purpose. The main goal of these approximations should be to exploit the information extracted from the set of promising solutions, allowing the EDA to explore promising areas of the search space.

Acknowledgements. This work has been partially supported by TIN2010-20900-C04-04, TIN2010-14931, Consolider Ingenio 2010-CSD2007-00018, Cajal Blue Brain projects (Spanish Ministry of Science and Innovation), the Saiotek and Research Groups 2007-2012 (IT-242-07) programs (Basque Government).

References

1. Agarwal, P., Procopiuc, C.: Exact and approximation algorithms for clustering. Algorithmica 33(2), 201–226 (2002)
2. Ahn, C.W., Ramakrishna, R.S., Goldberg, D.E.: Real-Coded Bayesian Optimization Algorithm: Bringing the Strength of BOA into the Continuous World. In: Deb, K., et al. (eds.) GECCO 2004. LNCS, vol. 3102, pp. 840–851. Springer, Heidelberg (2004)
3. Bengoetxea, E., Miquélez, T., Larrañaga, P., Lozano, J.A.: Experimental results in function optimization with EDAs in continuous domain. In: Larrañaga, P., Lozano, J.A. (eds.) Estimation of Distribution Algorithms. A New Tool for Evolutionary Computation, pp. 177–190. Kluwer Academic Publishers, Boston (2002)

4. Bosman, P.A.N., Thierens, D.: Expanding from Discrete to Continuous Estimation of Distribution Algorithms: The IDEA. In: Deb, K., Rudolph, G., Lutton, E., Merelo, J.J., Schoenauer, M., Schwefel, H.-P., Yao, X. (eds.) PPSN 2000. LNCS, vol. 1917, pp. 767–776. Springer, Heidelberg (2000)
5. Bosman, P.A.N., Thierens, D.: Numerical optimization with real-valued estimation-of-distribution algorithms. In: Pelikan, M., Sastry, K., Cantú-Paz, E. (eds.) Scalable Optimization via Probabilistic Modeling: From Algorithms to Applications. SCI, pp. 91–120. Springer (2006)
6. Chen, B., Hu, J.: A novel clustering based niching EDA for protein folding. In: Proceedings of the World Congress on Nature & Biologically Inspired Computing, NaBIC 2009, pp. 748–753. IEEE (2010)
7. Dill, K.A.: Theory for the folding and stability of globular proteins. Biochemistry 24(6), 1501–1509 (1985)
8. Dong, W., Chen, T., Tino, P., Yao, X.: Scaling up estimation of distribution algorithms for continuous optimization. CoRR, abs/1111.2221 (2011)
9. Efron, B., Hastie, T., Johnstone, I., Tibshirani, R.: Least angle regression. The Annals of Statistics 32(2), 407–499 (2004)
10. Fossati, L., Lanzi, P., Sastry, K., Goldberg, D., Gomez, O.: A simple real-coded extended compact genetic algorithm. In: Proceedings of the IEEE Congress on Evolutionary Computation, CEC 2007, pp. 342–348. IEEE (2007)
11. Frey, B.J., Dueck, D.: Clustering by passing messages between data points. Science 315, 972–976 (2007)
12. Friedman, J., Hastie, T., Tibshirani, R.: Sparse inverse covariance estimation with the graphical LASSO. Biostatistics 9(3), 432–441 (2008)
13. Gallagher, M., Frean, M., Downs, T.: Real-valued evolutionary optimization using a flexible probability density estimator. In: Proceedings of the Genetic and Evolutionary Computation Conference, GECCO 1999, Orlando, FL, vol. I, pp. 840–846. Morgan Kaufmann Publishers, San Francisco (1999)
14. Geman, S., Geman, D.: Stochastic relaxation, Gibbs distributions, and Bayesian restoration of images. IEEE Transactions on Pattern Analysis and Machine Intelligence 6(6), 721–741 (1984)
15. Hammersley, J.M., Clifford, P.: Markov fields of finite graphs and lattice. Technical report, University of California-Berkeley (1968)
16. Harik, G.: Linkage learning via probabilistic modeling in the ECGA. IlliGAL Report 99010, University of Illinois at Urbana-Champaign, Illinois Genetic Algorithms Laboratory, Urbana, IL (1999)
17. Hartigan, J., Wong, M.: Algorithm AS 136: A K-means clustering algorithm. Applied Statistics 28(1), 100–108 (1979)
18. Henrion, M.: Propagating uncertainty in Bayesian networks by probabilistic logic sampling. In: Lemmer, J.F., Kanal, L.N. (eds.) Proceedings of the Second Annual Conference on Uncertainty in Artificial Intelligence, pp. 149–164. Elsevier (1988)
19. Hesterberg, T., Choi, N., Meier, L., Fraley, C.: Least angle and L1 penalized regression: A review. Statistics Surveys 2, 61–93 (2008)
20. Hoerl, A.E., Kennard, R.W.: Ridge regression: Applications to nonorthogonal problems. Technometrics 12(1), 69–82 (1970)
21. Höns, R.: Estimation of Distribution Algorithms and Minimum Relative Entropy. PhD thesis, University of Bonn, Bonn, Germany (2006)
22. Hsu, H.-P., Mehra, V., Grassberger, P.: Structure optimization in an off-lattice protein model. Physical Review E 68(2), 4 Pages, article number 037703 (2003)

23. Karshenas, H., Santana, R., Bielza, C., Larrañaga, P.: Regularized model learning in estimation of distribution algorithms for continuous optimization problems. Technical Report UPM-FI/DIA/2011-1, Department of Artificial Intelligence, Faculty of Informatics, Technical University of Madrid (January 2011)
24. Kern, S., Müller, S.D., Hansen, N., Büche, D., Ocenasek, J., Koumoutsakos, P.: Learning probability distributions in continuous evolutionary algorithms– A comparative review. Natural Computing 3(1), 77–112 (2004)
25. Koller, D., Friedman, N.: Probabilistic Graphical Models: Principles and Techniques. Adaptive Computation and Machine Learning. The MIT Press (August 2009)
26. Lanzi, P., Nichetti, L., Sastry, K., Voltini, D., Goldberg, D.: Real-coded extended compact genetic algorithm based on mixtures of models. Linkage in Evolutionary Computation, 335–358 (2008)
27. Larrañaga, P., Etxeberria, R., Lozano, J., Peña, J.: Optimization in continuous domains by learning and simulation of Gaussian networks. In: Wu, A. (ed.) Conference on Genetic and Evolutionary Computation (GECCO 2000) Workshop Program, pp. 201–204. Morgan Kaufmann (2000)
28. Larrañaga, P., Lozano, J. (eds.): Estimation of Distribution Algorithms: A New Tool for Evolutionary Computation. Kluwer Academic Publishers, Norwell (2001)
29. Lauritzen, S.L.: Graphical Models. Oxford Statistical Science Series, vol. 17. Clarendon Press, Oxford (1996)
30. Li, M., Goldberg, D., Sastry, K., Yu, T.: Real-coded ECGA for solving decomposable real-valued optimization problems. Linkage in Evolutionary Computation, 61–86 (2008)
31. Malagó, L., Matteo, M., Gabriele, V.: Introducing l1-regularized logistic regression in Markov networks based EDAs. In: Proceedings of the 2011 Congress on Evolutionary Computation, CEC 2011, pp. 1581–1588. IEEE (2011)
32. Meinshausen, N., Bühlmann, P.: High-dimensional graphs and variable selection with the LASSO. Annals of Statistics 34(3), 1436–1462 (2006)
33. Mendiburu, A., Santana, R., Lozano, J.A.: Introducing belief propagation in estimation of distribution algorithms: A parallel framework. Technical Report EHU-KAT-IK-11/07, Department of Computer Science and Artificial Intelligence, University of the Basque Country (October 2007)
34. Mühlenbein, H., Paaß, G.: From Recombination of Genes to the Estimation of Distributions I. Binary Parameters. In: Ebeling, W., Rechenberg, I., Voigt, H.-M., Schwefel, H.-P. (eds.) PPSN 1996. LNCS, vol. 1141, pp. 178–187. Springer, Heidelberg (1996)
35. Ochoa, A.: Opportunities for Expensive Optimization with Estimation of Distribution Algorithms. In: Tenne, Y., Goh, C.-K. (eds.) Computational Intel. in Expensive Opti. Prob. ALO, vol. 2, pp. 193–218. Springer, Heidelberg (2010)
36. Pelikan, M., Goldberg, D.E., Tsutsui, S.: Getting the best of both worlds: Discrete and continuous genetic and evolutionary algorithms in concert. Information Sciences 156(3-4), 147–171 (2003)
37. Ravikumar, P., Raskutti, G., Wainwright, M., Yu, B.: Model selection in Gaussian graphical models: High-dimensional consistency of l1-regularized MLE. Advances in Neural Information Processing Systems (NIPS) 21 (2008)
38. Rue, H., Held, L.: Gaussian Markov Random Fields: Theory and Applications. Monographs on Statistics and Applied Probability, vol. 104. Chapman & Hall, London (2005)
39. Santana, R.: A Markov Network Based Factorized Distribution Algorithm for Optimization. In: Lavrač, N., Gamberger, D., Todorovski, L., Blockeel, H. (eds.) ECML 2003. LNCS (LNAI), vol. 2837, pp. 337–348. Springer, Heidelberg (2003)

40. Santana, R., Karshenas, H., Bielza, C., Larrañaga, P.: Regularized k-order Markov models in EDAs. In: Proceedings of the 2011 Genetic and Evolutionary Computation Conference, GECCO 2011, Dublin, Ireland, pp. 593–600 (2011)
41. Santana, R., Larrañaga, P., Lozano, J.A.: Learning factorizations in estimation of distribution algorithms using affinity propagation. Evolutionary Computation 18(4), 515–546 (2010)
42. Schäfer, J., Strimmer, K.: A shrinkage approach to large-scale covariance matrix estimation and implications for functional genomics. Statistical Applications in Genetics and Molecular Biology 4(1), 32 (2005)
43. Shakya, S.: Markov random field modelling of genetic algorithms. Technical report, The Robert Gordon University, Aberdeen, UK (2004)
44. Shakya, S., McCall, J.: Optimization by estimation of distribution with DEUM framework based on Markov random fields. International Journal of Automation and Computing 4(3), 262–272 (2007)
45. Shakya, S., Santana, R.: A Markovianity based optimisation algorithm. Genetic Programming and Evolvable Machines (2011) (in press)
46. Speed, T., Kiiveri, H.: Gaussian Markov distributions over finite graphs. The Annals of Statistics 14(1), 138–150 (1986)
47. Stillinger, F., Head-Gordon, T., Hirshfeld, C.: Toy model for protein folding. Physical Review E 48, 1469–1477 (1993)
48. Tibshirani, R.: Regression shrinkage and selection via the lasso. Journal of the Royal Statistical Society. Series B (Methodological) 58(1), 267–288 (1996)
49. Tsutsui, S., Pelikan, M., Goldberg, D.E.: Evolutionary algorithm using marginal histogram in continuous domain. In: Optimization by Building and Using Probabilistic Models, OBUPM 2001, San Francisco, California, USA, pp. 230–233 (July 2001)
50. Witten, D.M., Tibshirani, R.: Covariance-regularized regression and classification for high dimensional problems. Journal of the Royal Statistical Society. Series B. Methodological 71(3), 615–636 (2009)
51. Yuan, M., Lin, Y.: Model selection and estimation in the Gaussian graphical model. Biometrika 94(1), 19–35 (2007)
52. Zou, H., Hastie, T.: Regularization and variable selection via the elastic net. Journal of the Royal Statistical Society: Series B (Statistical Methodology) 67(2), 301–320 (2005)

Chapter 11
Using Maximum Entropy and Generalized Belief Propagation in Estimation of Distribution Algorithms

Robin Höns

Abstract. EDAs work by sampling a population from a factorized distribution, like the Boltzmann distribution of an additively decomposable fitness function (ADF). In the Factorized Distribution Algorithm (FDA), a factorization is built from an ADF by choosing a subset of the factors. I present a new algorithm to merge factors into larger sets, allowing to account for all dependencies between the variables. Estimating larger subset distributions is more prone to sample noise, so the larger distribution can be estimated from the smaller ones with the Maximum Entropy method. Building an exact graphical model for sampling is often infeasible. E.g. in a 2-D grid the triangulated Markov network has linear clique size, thus exponentially large distributions. I explore ways to use loopy models and Generalized Belief Propagation in the context of EDA and optimization. The merging algorithm mentioned above can be combined with this.

11.1 Introduction

EDAs are population-based optimization algorithms. Most generally, in the selection stage they choose the fittest individuals from the population, and then – where the good old Genetic Algorithm used crossover and mutation to create a new population – they devise a probabilistic model of the search space and sample new points from it.

It started fifteen years ago with the Univariate Marginal Distribution Algorithm (UMDA) [22], where the model consists just of the univariate marginal probabilities of the variables, without any connections between them. Then, factorization systems were used to model these connections, resulting in the Factorized Distribution Algorithm (FDA) [19]. These factorization systems are almost identical to Bayesian networks.

Robin Höns
e-mail: robin@hoens.net

S. Shakya and R. Santana (Eds.): Markov Networks in Evolutionary Computation, ALO 14, pp. 175–190.
springerlink.com

These algorithms assume that the dependency structure is known. Of course, in general this is not the case. But whereas most efforts in EDA since then have consisted in learning a dependency structure from a population, in this chapter we assume that the structure is known.

Information about the structure can be given in different ways. FDA works with a factorization system, i. e. a system of variable sets. Therefore, if a Markov network is given, it has to be transformed into a factorization system. Originally, FDA uses a subset of the cliques of the Markov network, and disregards some connections. The correct solution would be to triangulate the network, but this usually results in infeasibly large cliques.

Section 11.3 presents an algorithm which merges the cliques of the Markov network in order to capture more or even all dependencies, but avoiding complete triangulation.

Merging cliques means larger cliques, of course. And since FDA estimates distributions on the cliques, the estimation from a given population size gets worse with larger clique size. Section 11.3.4 presents a way to estimate these distributions using the Maximum Entropy principle.

Another possibility, short of triangulating the Markov network, is to create a loopy structure and use generalized belief propagation (GBP) on it. This approach is explored in section 11.4. First, the loopy structure (the region graph) is presented, then follows the GBP algorithm and a way to use it for optimization, and finally some numerical results.

Finally, section 11.5 concludes the chapter and gives some possibilities of further work.

11.2 The Factorized Distribution Algorithm

Let there be given a fitness function $f : \{0,1\}^n \to \mathbb{R}$. An *additive decomposition* $\mathfrak{S}$ of f is a system of index sets $s_1,\ldots,s_m \subseteq \{1,\ldots,n\}$ and a set of functions $f_1(\mathbf{x}_{s_1}),\ldots,f_m(\mathbf{x}_{s_m})$, where $\mathbf{x}_{s_i}$ denotes the subvector of $\mathbf{x} \in \{0,1\}^n$ indexed by s_i, such that

$$f(\mathbf{x}) = \sum_{i=1}^{m} f_i(\mathbf{x}_{s_i}) \tag{11.1}$$

If there exists an additive decomposition of f, we call f an *additively decomposable function (ADF)*.

For an additive decomposition, we define $d_i := \bigcup_{j=1}^{i} s_j$, $b_i := s_i \setminus d_{i-1}$, and $c_i := s_i \cap d_{i-1}$ (with $d_0 := \emptyset$). An additive decomposition $\mathfrak{S}$ is a *factorization system* if $d_m = \{1,\ldots,n\}$, so that in the end all variables are contained, and all $b_i \neq \emptyset$. The *Boltzmann distribution* for f a probability distribution

$$p(\mathbf{x}) = \frac{1}{Z} e^{\beta f(\mathbf{x})} \tag{11.2}$$

where β is called the *inverse temperature* and Z the *partition function* for normalization.

Using the factorization system, we can define the distribution

$$\tilde{p}(\mathbf{x}) = \prod_{i=1}^{m} p(\mathbf{x}_{b_i}|\mathbf{x}_{c_i}) \tag{11.3}$$

It can be shown that for all i, $\tilde{p}(\mathbf{x}_{b_i}|\mathbf{x}_{c_i}) = p(\mathbf{x}_{b_i}|\mathbf{x}_{c_i})$ [3, p. 24].

But $\tilde{p} = p$ is only valid if the factorization system fulfills the Running Intersection Property (RIP) [13]:

$$\forall i \geq 2 \quad \exists j < i: \quad c_i \subseteq s_j \tag{11.4}$$

This means that all the variable sets on which new variables are conditioned must be contained in a previous subset.

Given an ADF $f(\mathbf{x})$, whose factorization satisfies the RIP, the maximum of $f(\mathbf{x})$ can be found in polynomial time (provided that the size of the subsets is at most logarithmic in n) by dynamic programming. Algorithms for finding the optimum can be found in [1, 10].

So, it is no surprise that in this case $\tilde{p} = p$ holds. This result is called the *Factorization Theorem*. The proof can be found in [14] or [3, p. 29].

This theorem induces the Factorized Distribution Algorithm (FDA) [19, 21] (Alg. 1).

Algorithm 1. ***FDA* – Factorized Distribution Algorithm**

---

1 Calculate b_i and c_i from the decomposition of the function.
2 *Let* t 1. Generate an initial population with N individuals from the uniform distribution.
3 **do** {
4 Perform selection
5 Estimate the conditional probabilities $p(\mathbf{x}_{b_i}|\mathbf{x}_{c_i},t)$ from the selected points.
6 Generate new points according to $p(\mathbf{x},t+1) = \prod_{i=1}^{m} p(\mathbf{x}_{b_i}|\mathbf{x}_{c_i},t)$.
7 $t \Leftarrow t+1$.
8 } **until** stopping criterion reached

A difficulty of *FDA* is the choice of the population size. If it is too small, the estimate of $p(\mathbf{x}_{b_i}|\mathbf{x}_{c_i},t)$ will be bad. On the other hand, a large population slows the algorithm down, especially if fitness function evaluation is expensive.

11.3 Finding a Factorization System

FDA needs a factorization of f. Given an ADF f, we need two things to turn the additive decomposition into a factorization system:

1. An ordering of the subsets and
2. A way to handle $b_i = \emptyset$.

Given an additive decomposition $\mathfrak{S} = \{s_1, \ldots, s_m\}$, it is often not possible to order the s_i in a factorization system, because this results in a $b_i = \emptyset$; that is, a set whose variables are all already contained in previous sets. Take for example the decomposition $(\{1,2\},\{1,3\},\{2,3\})$. No matter which ordering we choose, the third set will be fully contained in the union of the previous sets.

11.3.1 Subfunction Choice

In this case, it is in general not possible to sample from the system in a consistent way. One possibility to solve this problem is to choose only a subset of the s_i and disregard the others; in our example, we can use the factorization $(\{1,2\},\{1,3\})$, so $p(x_1,x_2,x_3) = p(x_1,x_2)p(x_3|x_1)$.

To choose a subset $\tilde{\mathfrak{S}} = \{\tilde{s}_j | 1 \le j \le \tilde{m}\} \subseteq \mathfrak{S}$ which is a factorization system, it is desirable to use as many connections between the variables as possible. This is implemented in the heuristics of Alg. 2 [20].

Algorithm 2. **Subfunction Choice**

1. Choose the first set $\tilde{s}_1$ randomly among the s_i.
2. Then, choose among the s_i the set that has the maximal overlap with $\tilde{d}_{j-1}$ (the history of variables added so far). If there is more than one possible set, choose the smallest one.
3. Add this set as $\tilde{s}_j$. $j \Leftarrow j+1$
4. Repeat this until the whole set $\{1,\ldots,n\}$ is covered.

11.3.2 Subfunction Merge

Another possibility is, instead of leaving out some subfunctions, merging them. By merging subfunctions, it is possible to create a factorization system that accounts for all connections and even complies with the running intersection property. Actually, this is obvious, since the trivial *complete* merge of all subfunctions has this property. This is of course useless, because such big sets of variables make the algorithms exponentially costly.

We now present a heuristic algorithm that merges subfunctions in order to use all dependencies, but minimizing the number of merges.

The idea of Alg. 3 is that each new variable is added in a set with the previous variables on which it depends. However, if another variable depends on a superset of variables, the two are merged and added together.

Algorithm 3. **Subfunction Merger**

```
1  𝒮 ⇐ {s_1,...,s_m}
2  j ⇐ 1
3  while d̃_j ≠ {1,...,n} do {
4      Choose an s_i ∈ 𝒮 to be added, the same way as in Alg. 2
5      𝒮 ⇐ 𝒮 \ {s_i}
6      Let the new variable indices in s_i be b_i = {k_1,...,k_l}
7      for λ = 1 to l do {
8          Let δ_λ be all variables in d̃_{j-1} on which k_λ depends
9      }
10     for λ = 1 to l do {
11         if exists λ' ≠ λ with δ_λ ⊆ δ_λ' and not exists λ'' with δ_λ' ⊂ δ_λ''
12             δ_λ' ⇐ δ_λ' ∪ {k_λ}
13             Mark k_λ superfluous
14     }
15     for λ = 1 to l do {
16         if not k_λ superfluous
17             s̃_j ⇐ δ_λ ∪ {k_1,...,k_λ}
18             while |s̃_j| > maximal cluster size do {
19                 Choose randomly an index k from c̃_j
20                 Remove k from s̃_j
21             }
22             j ⇐ j+1
23     }
24 }
```

Every time we add a new subset s_i, we need to consider the new variables b_i to be added to the factorization system. Let the variable indices in b_i be $b_i = \{k_1, \ldots, k_l\}$. For each $\lambda \in \{1, \ldots, l\}$, the indices of all previous variables on which x_{k_λ} depends are stored in δ_λ. However, if there is a λ' with $\delta_\lambda \subseteq \delta_{\lambda'}$, it suffices to add only one set $\{k_\lambda, k_{\lambda'}\} \cup \delta_{\lambda'}$. Therefore k_λ is marked superfluous and, instead, added to $\delta_{\lambda'}$ in order to be added with this subset.

For densely connected problems, it can occur that the sizes of the sets $|\tilde{s}_j|$ become too large. Therefore we introduce a maximal size of the subsets. If this is exceeded, we remove variables from $\tilde{s}_j$ randomly. We remove only dependencies on previous variables, i. e. indices from $\tilde{c}_j$, not indices of new variables from $\tilde{b}_j$.

Removing variables randomly is probably not the best choice. Other schemes could be conceived, e. g. deviation from linear regression in the fitness. Note however that unlike Alg. 3, such a scheme requires evaluation of the fitness function.

11.3.3 An Important Example: Pentavariate Factorization of the 2-D Grid

To make the algorithm clearer, here is an example. A very important dependency structure is the 2-D grid. An example function is the *Deceptive 4-Grid* with

$$f_{\text{Dec4}}(u := x_{i,j} + x_{i+1,j} + x_{i,j+1} + x_{i+1,j+1}) = \begin{cases} 3-u & \Longleftrightarrow u < 4 \\ 4 & \Longleftrightarrow u = 4 \end{cases} \quad (11.5)$$

and the fitness function being the sum of these subfunctions for all 2×2 blocks, with overlapping blocks:

$$F_{\text{Dec4}}(\mathbf{x}) = \sum_{i=1}^{m-1} \sum_{j=1}^{m-1} f_{\text{Dec4}}(x_{i,j} + x_{i+1,j} + x_{i,j+1} + x_{i+1,j+1}) \quad (11.6)$$

An exact factorization of the grid would result in cliques of linear size [15] and therefore in exponentially large marginal distributions.

Alg. 2 results in the factorization system depicted on the left side of Fig. 11.1 [14, Sect. 4.2.6]. It is possible to sample points from it, using the probability distribution

$$p_{\text{Fact4}}(\mathbf{x}) = p(x_1,x_2,x_5,x_6)p(x_3,x_7|x_2,x_6)p(x_4,x_8|x_3,x_7) \\ p(x_9,x_{10}|x_5,x_6)p(x_{11}|x_6,x_7,x_{10})p(x_{12}|x_7,x_8,x_{11}) \quad (11.7)$$

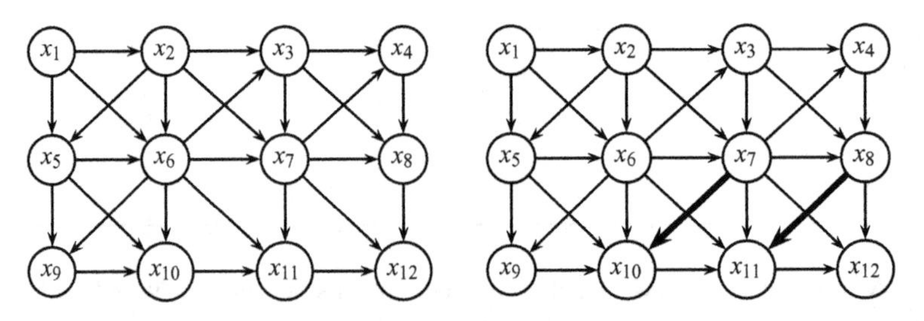

Fig. 11.1 The Bayesian network for the example 4×4 grid and the pentavariate factorization. The fat arrows are the added dependencies.

Using the merge algorithm (Alg. 3) we get the following factorization:

$$p_{\text{Fact5}}(\mathbf{x}) = p(x_1,x_2,x_5,x_6)p(x_3,x_7|x_2,x_6)p(x_4,x_8|x_3,x_7) \\ p(x_9,x_{10}|x_5,x_6,x_7)p(x_{11}|x_6,x_7,x_8,x_{10})p(x_{12}|x_7,x_8,x_{11}) \quad (11.8)$$

Since the marginals contain up to five variables, we call this factorization the *pentavariate* factorization. It is depicted on the right side of Fig. 11.1.

11.3.4 With Maximum Entropy

Merging subfunctions has a severe disadvantage: By using larger marginal distributions, more connections between nodes are captured, but the number of degrees of freedom grows exponentially in the size of the marginal. Thus the estimate is more sensitive to noise in the data.

However, there is a remedy. Consider one of the pentavariate subsets, say $x_5, x_6, x_7, x_9, x_{10}$. Remember that a Markov network codes information about conditional independencies. In this case, x_7 is conditionally independent of x_5, x_9, given x_6, x_{10}. So we can separate the set in two parts: x_5, x_6, x_9, x_{10} and x_6, x_7, x_{10}. So we got rid of the pentavariate set and just have two sets of four and three variables, thus less degrees of freedom.

For the 2-D grid, the gain is not very large; there are other structures where the difference is larger. E. g. on an analogous 3-D grid, the method reduces the marginals from $2^{11} = 2048$ to $2^8 + 2^7 = 384$ degrees of freedom.

The idea is to estimate the smaller marginals from the population and combine them to form the large marginal.

But how should the large marginal be formed from the smaller ones? We use the distribution with *Maximum Entropy*, given the marginals. Unfortunately there is no space to motivate this choice, tell the fascinating history of this principle and present the interesting proofs. But I recommend the very interesting article [7] or [3, Chapter 4].

The maximum entropy distribution, given some marginals, can be calculated using *Iterative Proportional Fitting* (IPF) [2, 5]. The complete proof that IPF converges to the maximum entropy distribution can be found in [3, Sect. 4.2.6].

IPF computes iteratively a distribution $q_\tau(\mathbf{x})$ from the given marginals $p_k(\mathbf{x}_{s_k})$, where $\tau = 0, 1, 2, \ldots$ is the iteration index. Most commonly, $q_{\tau=0}$ is the uniform distribution. The update formula is

$$q_{\tau+1}(\mathbf{x}) = q_\tau(\mathbf{x}) \frac{p_k(\mathbf{x}_{s_k})}{q_\tau(\mathbf{x}_{s_k})} \qquad \text{with } k = ((\tau - 1) \mod K) + 1 \tag{11.9}$$

IPF takes exponential time and space, since q_τ is exponential. But using graphical models, e. g. a junction tree, the complexity can be reduced [9].

To make the resulting algorithm clearer, we put together all the parts described so far and present the complete pseudocode in Alg. 4.

Algorithm 4. ***MEFDA* – Maximum Entropy *FDA***

1 Using Alg. 3, construct a merged factorization system $\tilde{\mathfrak{S}}$ from the given additive decomposition $\mathfrak{S}$
2 $t \Leftarrow 1$. Generate an initial population with N individuals from the uniform distribution.
3 **do** {
4 Perform selection
5 Estimate the marginal probabilities $p(\mathbf{x}_{s_i}, t)$ from the selected points.
6 Calculate all $p(\mathbf{x}_{\tilde{s}_j}, t)$ by IPF using $\mathfrak{S}_{\mathrm{IPF},j}$
7 Generate new points according to $p(\mathbf{x}, t+1) = \prod_{j=1}^{m} p(\mathbf{x}_{\tilde{b}_j} | \mathbf{x}_{\tilde{c}_j}, t)$.
8 $t \Leftarrow t+1$.
9 } **until** stopping criterion reached

11.3.5 Numerical Results

In this section two new variants of the *FDA* algorithm were presented:

- *FDA* with merging of subfunctions using Alg. 3 (instead of picking a subset of the factorization sets using Alg. 2) and
- *MEFDA* (Alg. 4), merging subfunctions like above, but using IPF to construct the larger marginal distributions.

Now we compare these algorithms with conventional *FDA* (Alg. 1). Their results on the deceptive grid function (11.6) are depicted in Table 11.1.

We are mostly interested in the minimal population size for which the algorithms have a large success rate (above 95 %). It can be noticed that both new algorithms perform better than *FDA*. They manage to reduce the population size impressively. Also, the number of required generations until convergence is reduced. Since the number of function evaluations is proportional to the population size and the number of generations, the methods allow large savings in the number of function evaluations.

FDA with merge and *MEFDA* need a comparable number of generations until convergence. *MEFDA* needs slightly more generations, but manages to decrease the required population size even more than *FDA* with subfunction merge. The price to pay for this is the additional overhead of the IPF procedures, but it is decent for not too large set sizes, and furthermore, IPF does not need any fitness function evaluations. So these methods are particularly interesting for problems in which evaluation of the fitness function is expensive.

The difference between *FDA* with merge and *MEFDA* becomes smaller with increasing grid size. It can be seen in Table 11.1 that for the 25×25 grid the success rates are almost the same. The reason for this is that both algorithms need a larger population for large grids. Large populations lead to better estimations of the marginal distributions. Therefore, the large marginals created by the merging

Table 11.1 Results of the merging and Maximum Entropy methods on the deceptive grid (11.6) of different grid sizes and population sizes. *SR* gives the success rate (number of successful runs), *Gen* the average number of generations until success and *SD* its standard deviation. Parameters: 100 runs each. Truncation Selection with $\tau = 0.3$. Maximal number of generations is 20 for the 10×10 and 15×15 grids, 30 for the bigger grids.

Grid size	Pop. size	*FDA* SR Gen ± SD	*FDA* with merge SR Gen ± SD	*MEFDA* SR Gen ± SD
10×10 = 100	700	79 8.58 ± 0.810	70 7.81 ± 0.748	98 7.58 ± 0.745
	900	92 8.54 ± 0.895	91 7.44 ± 0.733	100 7.33 ± 0.667
	1000	95 8.36 ± 0.757	99 7.45 ± 0.689	100 7.22 ± 0.786
15×15 = 225	1200	34 15.38 ± 0.853	64 13.17 ± 1.121	96 13.15 ± 0.754
	1600	56 14.68 ± 1.081	96 12.57 ± 0.692	98 12.83 ± 0.643
	2000	71 14.70 ± 0.852	98 12.10 ± 0.601	100 12.48 ± 0.643
	2700	89 14.29 ± 0.920	100 11.92 ± 0.580	100 12.07 ± 0.537
20×20 = 400	1900	8 21.88 ± 0.641	64 18.25 ± 0.836	97 18.75 ± 0.902
	2000	5 21.20 ± 1.095	77 17.94 ± 0.767	97 18.43 ± 0.900
	2400	17 21.65 ± 0.996	95 17.65 ± 0.796	100 18.28 ± 0.697
	3000	31 21.10 ± 1.106	99 17.16 ± 0.584	100 18.05 ± 0.609
25×25 = 625	3300	2 27.50 ± 0.707	94 22.61 ± 0.722	99 24.31 ± 0.804
	3600	3 28.33 ± 0.578	99 22.44 ± 0.610	99 24.11 ± 0.754

algorithm can be estimated well from the population, and the precision gain of IPF is lost.

11.4 Loopy Belief Propagation

Another very interesting topic of research is the generalization of graphical models. For inference, a Markov network can be converted into a junction tree [4, 8] by the well-known steps:

1. Triangulate the Markov network
2. Find the cliques
3. Build the junction tree: Each clique is turned into a node, the edges (separators) between two nodes are marked with the intersections of the two cliques.

Triangulation often results in a too densely connected graph, thus too large cliques. E. g. for a grid, the cliques would consist of two neighboring rows [15]. So what happens if we leave this step out?

11.4.1 The Region Graph

Without triangulation, the junction tree turns into a loopy structure, called *region graph* [25] or *poset* [16]. One way to build it is to find the cliques, then find the intersections of these cliques, then the intersections between these sets, etc. This algorithm is called *cluster variation method* (CVM) [11].

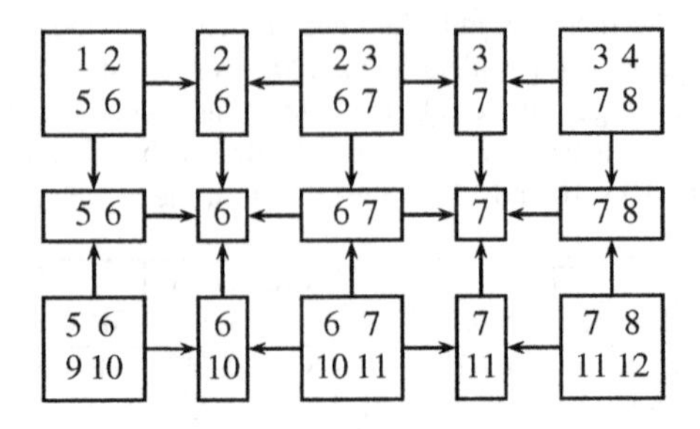

Fig. 11.2 The region graph for a rectangular grid, using 2×2 blocks

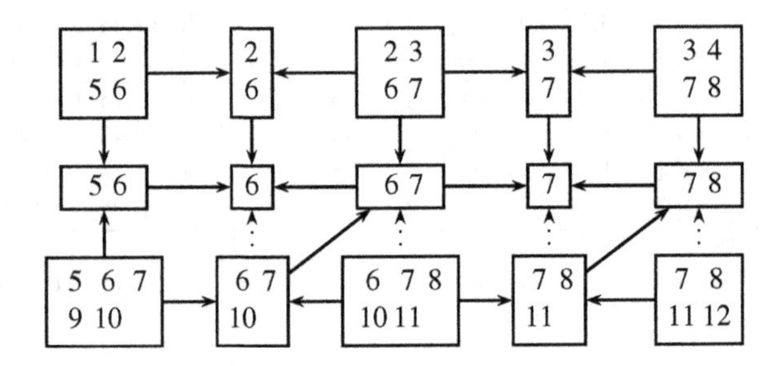

Fig. 11.3 The region graph for the pentavariate factorization

Another possibility is to create a large region for each clique and a region for each variable; then connect the variable node for X_i with each clique which contains X_i. This special structure is also called *factor graph* [12].

Fig. 11.2 and 11.3 give the region graphs for the 2-D grid, 4-variate and 5-variate factorization, created by the CVM.

The dotted edges in Fig. 11.3 are optional; they connect grandparent and grandchild, so they can be left out. This doesn't change the result of the algorithm, only the dynamics. Including them slightly speeds up the algorithm.

11.4.2 Belief Propagation on the Region Graph

Belief propagation on the junction tree [4] can be generalized for this structure [25]. The algorithm minimizes the *Gibbs free energy*, a notion from statistical physics. Unfortunately I cannot present the complete connection to statistical physics here, so once again I refer the interested reader to [25] or [3].

Here we just give the result of the algorithm. For each region, the *local belief* q_R is defined as

$$q_R(\mathbf{x}_R) \propto \prod_{f_i \in F_R} e^{\beta f_i(\mathbf{x}_{s_i})} \prod_{(P,C) \in M(R)} m_{P \to C}(\mathbf{x}_C) \tag{11.10}$$

where F_R is the set of subfunctions f_i of the ADF defined on the region R and $m_{P \to C}(\mathbf{x}_C)$ is a set of *messages* between parent and child, updated in each step of the algorithm. These messages are used to achieve consistency between neighbouring regions.

$M(R)$ is the set of all *relevant* edges for the region R: These are the edges leading to R or a descendent of R, coming from a non-descendent.

The algorithm for updating the messages, thus learning the distributions q_R and achieving consistence between them, can be found in [3].

11.4.3 Optimization Using Loopy Models

Our goal is to use the methods described so far for optimization. GBP provides us with a set of marginal distributions for the regions, all consistent with each other. But it does not give us a global distribution from which we could sample. One possibility is to use Gibbs sampling [24]. However, here we will pursue the idea to combine this technique with the factorizations of *FDA* in order to sample points.

This section presents the new method which combines factorization systems with GBP to devise the *BKDA* (Bethe Kikuchi Distribution Algorithm) [18]. It consists of the following basic steps:

1. Construct a factorization system from the ADF, with or without merging.
2. Construct a region graph from the ADF and the factorization system, using CVM.
3. Perform GBP on the region graph.
4. Calculate the marginals of the factorization from the result of GBP.
5. Sample points from this distribution.

In the literature the inverse temperature β in (11.10) is mostly disregarded. But in our optimization context, β plays a vital role. If it is too low, the probability of the maximum is low, too. However, for too high values of β, convergence of the algorithm becomes more and more difficult. It needs more iteration steps to converge. The beliefs (11.10) and messages depend exponentially on β, so for very large β they become numerically unstable.

It is possible to use a kind of "cooling schedule". This means to start with a low β and then use the result of GBP as the starting point for the next GBP run with a higher value of β. This could improve performance, but not the result, since the fixed points of GBP are independent from the starting points.

11.4.4 Numerical Results: Deceptive Grid

The results of BKDA on the deceptive grid problem are presented in Table 11.2. Given are the probabilities of sampling the optimum and the steps until convergence, using a 4-grid region graph, a 5-grid region graph, and for comparison the exact solution using a junction tree.

Table 11.2 Results of *BKDA* on Deceptive-4-Grid. Depicted is the probability of sampling the known optimum using the factorization and the number of steps until convergence, with the 4-grid and pentavariate grid model, and the exact junction tree probability. $\beta = 1$.

Size	4-Grid	Penta-Grid	Junction tree
16	0.918358 31	0.918454 32	0.91842
25	0.918748 47	0.918809 44	0.918813
36	0.917477 37	0.917537 38	0.91754
49	0.916179 33	0.91624 36	0.916243
64	0.914882 32	0.914943 36	0.914946
81	0.913587 32	0.913648 36	0.913651
100	0.912294 32	0.912354 36	0.912357
121	0.911002 32	0.911062 36	0.911065

It can be noted that the probability of the optimum is almost independent of the grid size. The problem is very well-behaved. Both factorizations can reproduce the probability of the optimum almost exactly; the pentavariate factorization is slightly more accurate, but requires slightly more iterations. The number of iterations is independent of the grid size.

11.4.5 Numerical Results: Ising Spinglass

Next, we apply these algorithms on the Ising spinglass problem [6, 17, 23]. It is defined as finding the "spin" vector $\mathbf{s} = (s_1, \ldots, s_n)$ minimizing the Hamiltonian energy

$$E(\mathbf{s}) = - \sum_{1 \le \nu < \mu \le n} J_{\mu,\nu} s_\mu s_\nu \tag{11.11}$$

where the variables $s_\nu \in \{-1, +1\}$. They are arranged on a 2-D grid, so the *coupling constants* $J_{\mu,\nu}$ are only non-zero for neighbours on the grid. Note that the problem is symmetric, so inverting each spin does not change the energy. Therefore, a spin glass always has at least two global optima (ground states).

Due to space restrictions, we only present the results on ten random instances of an 10×10 grid[1]. Table 11.3 gives the probability of sampling the optimum, after the distributions have been learned by *BKDA*.

The fitness landscape is much more complicated than for the deceptive grid. Therefore, the approximation is not exact, and the probability of the optimum is quite different from the exact junction tree result. Note that in all instances the pentavariate grid performs better than the 4-grid.

It can be seen that the probability depends heavily on β. If the approximation was exact, it should converge against 0.5 for $\beta \to \infty$ (since there are two optima). But Fig. 11.4 shows that life is not that simple.

[1] The used instances are available at http://hoens.net/robin/ising for download. Their solutions have been verified by the Cologne spin glass server http://www.informatik.uni-koeln.de/spinglass.

Table 11.3 Results of *BKDA* on Ising instances of size 10×10. Depicted is the probability of sampling the known optimum using the factorization, with the 4-grid and pentavariate grid model, and the exact junction tree probability. *: Instance 6 uses a different procedure (CCCP) due to convergence problems.

	$\beta = 1$			$\beta = 10$		
Seed	4-Grid	Penta-Grid	J. Tree	4-Grid	Penta-Grid	J. Tree
1	1.56227e-13	5.53463e-13	1.25128e-12	0.0158715	0.0143937	0.0598039
2	1.33401e-12	1.41467e-11	4.8858e-11	0.00577445	0.0129105	0.0988492
3	5.55311e-12	2.56564e-11	5.95572e-11	0.133502	0.134356	0.144596
4	8.73883e-12	3.66898e-11	7.73724e-11	0.0114817	0.0221649	0.107321
5	7.42522e-12	2.4684e-11	5.21087e-11	0.0281239	0.0358553	0.0760384
6	6.29962e-12	3.10583e-11	9.27718e-11	0.0791284*	0.0794609*	0.21017
7	2.75456e-11	1.16145e-10	2.12433e-10	0.196065	0.220916	0.343579
8	3.13778e-12	1.33932e-11	3.28464e-11	0.000738162	0.000740625	0.187813
9	5.05162e-13	4.72522e-12	1.00587e-11	3.60448e-05	0.000253353	0.106966
10	3.26034e-11	2.7924e-10	4.52043e-10	0.000203547	0.0877621	0.219027

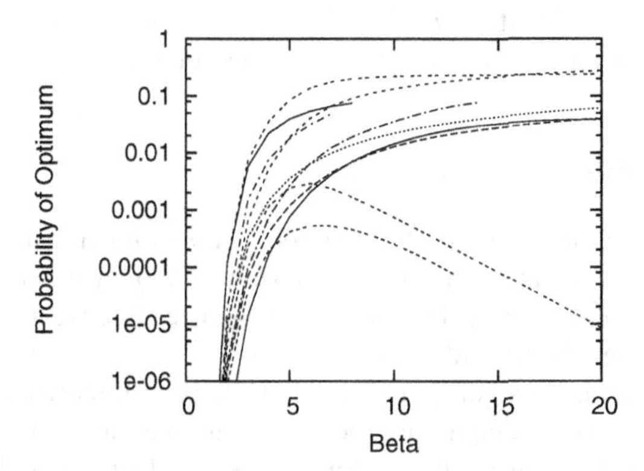

Fig. 11.4 Probability of the optimum for the Ising instances for varying β. Pentavariate grid, $\alpha = 0.5$.

For some problems, the probability stays quite high with rising β, but for some it declines again. And some lines could not be drawn completely, because the algorithm wouldn't converge any more due to numerical instability. [3, Sect. 6.6] presents a way to deal with such instabilities using a different update scheme, the Concave Convex Procedure (CCCP).

For comparison, the results of *FDA* and *MEFDA* on the same instances are given in Table 11.4. We see that probably *FDA* with the pentavariate grid performs best on these problems. (On smaller problems, *MEFDA* is better, but just like in Sect. 11.3.5, on larger populations *MEFDA* loses its advantage.)

Comparing *FDA* with *BKDA*, we see that *BKDA*, after one run of GBP, allows to find the optimum with a sample size of $N = 1000$ (except for instance 9, which would rather need about $N = 10000$ samples). Conversely, on instance 2 *FDA* is

Table 11.4 Results of *FDA* and *MEFDA* on the Ising instances. SR gives the successful runs out of 100, Gen the average generations until success and Sdv the standard deviation of the generations until success. 20 runs, population size $N = 1000$, truncation threshold $\tau = 0.3$.

	FDA 4-grid			*MEFDA* 4-grid			*FDA* 5-grid			*MEFDA* 5-grid		
Seed	SR	Gen	Sdv	SR	Gen	Sdv	SR	Gen	Sdv	SR	Gen	Sdv
1	17	14.35	1.27	19	14.53	1.50	16	14.00	1.37	15	14.40	1.18
2	10	15.10	2.38	9	15.22	1.64	13	13.46	1.33	9	15.33	1.80
3	12	13.50	1.51	13	13.31	1.18	13	12.15	0.99	13	13.08	1.26
4	14	13.29	1.27	12	13.83	1.40	13	12.62	1.66	10	13.70	0.82
5	20	13.05	1.43	20	13.50	0.95	20	12.55	1.05	20	13.45	1.50
6	15	12.67	0.82	10	14.10	1.91	19	11.89	0.88	17	14.24	1.89
7	19	12.05	1.08	20	12.25	0.79	18	11.28	1.18	19	12.47	0.96
8	10	12.60	1.26	13	12.62	1.04	14	12.29	0.91	16	12.88	1.20
9	4	14.75	0.50	6	14.67	1.51	18	12.72	1.02	6	14.33	0.82
10	14	13.21	0.97	15	12.87	1.60	16	11.81	1.28	14	13.14	1.23

not very successful, whereas *BKDA* has no particular problem with this one. But we should keep in mind that the *FDA* algorithms need much more fitness function evaluations (population size times number of generations).

11.5 Conclusion

This chapter has presented possibilities to improve the estimation of probability distributions using the principle of maximum entropy and loopy models, and their application in the context of EDA. They have proven to be effective, but more research in this area will surely be fruitful.

The subfunction merge algorithm allows FDA to exploit more dependencies between the variables. If the cliques are too large, variables are removed by random. This is probably not the best solution. Several other schemes could be tried, e. g. deviance from linear regression in the fitness value. Maybe using different models for different parts of the population or dropping different dependencies between the generations could be worthwhile. Then, perhaps the maximum entropy approach could be tried with EDAs which learn the structure from the population. This would require additional information about the variable dependencies. Another possible alternative to consider is learning two networks $\mathfrak{S}$ and $\tilde{\mathfrak{S}}$ with different structure penalty parameters, estimate a distribution of $\mathfrak{S}$ from the population, use these marginals for IPF on $\tilde{\mathfrak{S}}$ and sample the new population using $\tilde{\mathfrak{S}}$.

For the loopy belief propagation approach there are various ways to build a region graph, apart from the mentioned cluster variation method. E. g. the factor graph is a special case of the region graph. There are also algorithms for learning a structure from a population. It could be investigated how GBP performs on these structures, and if it works well for optimization.

The most interesting lesson to be learned is that interdisciplinary research, like in this case bringing together information theory, statistical physics and probability calculus, can lead to new insights. A great advantage is that it is easier to analyze the algorithm if there is a sound theoretical foundation, and all these fields bring with them a powerful set of analyzing tools.

References

1. Bertelè, U., Brioschi, F.: Nonserial Dynamic Programming. Academic Press, New York (1972)
2. Deming, W.E., Stephan, F.F.: On a least square adjustment of a sampled frequency table when the expected marginal totals are known. Ann. Math. Statist. 11, 427–444 (1940)
3. Höns, R.: Estimation of Distribution Algorithms and Minimum Relative Entropy. PhD thesis, University of Bonn, Germany (2006), URN: urn:nbn:de:hbz:5N-07741, `http://hss.ulb.uni-bonn.de/2006/0774/0774.htm`
4. Huang, C., Darwiche, A.: Inference in belief networks: A procedural guide. International Journal of Approximate Reasoning 15(3), 225–263 (1996)
5. Ireland, C.T., Kullback, S.: Contingency tables with given marginals. Biometrika (1968)
6. Ising, E.: Beitrag zur Theorie des Ferromagnetismus. Zeitschr. f. Physik 31, 253–258 (1925)
7. Jaynes, E.T.: Where do we stand on maximum entropy? In: Levine, R.D., Tribus, M. (eds.) The Maximum Entropy Formalism. MIT Press, Cambridge (1978)
8. Jensen, F.V., Jensen, F.: Optimal junction trees. In: Proceedings of the 10th Conference on Uncertainty in Artificial Intelligence, Seattle, pp. 360–366 (1994)
9. Jiroušek, R., Přeučil, S.: On the effective implementation of the iterative proportional fitting procedure. Computational Statistics & Data Analysis 19, 177–189 (1995)
10. Jordan, M.I. (ed.): Learning in Graphical Models. MIT Press, Cambridge (1999)
11. Kikuchi, R.: A theory of cooperative phenomena. Phys. Review 115, 988–1003 (1951)
12. Kschischang, F.R., Frey, B.J., Loeliger, H.-A.: Factor graphs and the sum-product algorithm. IEEE Trans. on Information Theory 47(2), 498–519 (2001)
13. Lauritzen, S.L.: Graphical Models. Clarendon Press, Oxford (1996)
14. Mahnig, T.: Populationsbasierte Optimierung durch das Lernen von Interaktionen mit Bayes'schen Netzen. PhD thesis, Universität Bonn (2001)
15. Martelli, A., Montanari, U.: Nonserial dynamic programming: On the optimal strategy of variable elimination for the rectangular lattice. J. Math. Anal. Appl. 40, 226–242 (1972)
16. McEliece, R.J., Yildirim, M.: Belief propagation on partially ordered sets. In: Proceedings of the 15th Internatonal Symposium on Mathematical Theory of Networks and Systems, MTNS 2002 (2002)
17. Mézard, M., Parisi, G., Virasoro, M.A.: Spin glass theory and beyond. World Scientific, Singapore (1987)
18. Mühlenbein, H., Höns, R.: The estimation of distributions and the minimum relative entropy principle. Evolutionary Computation 13(1), 1–27 (2005)
19. Mühlenbein, H., Mahnig, T.: Convergence theory and applications of the factorized distribution algorithm. Journal of Computing and Information Technology 7(1), 19–32 (1998)
20. Mühlenbein, H., Mahnig, T.: FDA - a scalable evolutionary algorithm for the optimization of additively decomposed functions. Evolutionary Computation 7(4), 353–376 (1999)
21. Mühlenbein, H., Mahnig, T., Ochoa, A.: Schemata, distributions and graphical models in evolutionary optimization. Journal of Heuristics 5(2), 213–247 (1999)

22. Mühlenbein, H., Paaß, G.: From Recombination of Genes to the Estimation of Distributions I. Binary Parameters. In: Ebeling, W., Rechenberg, I., Voigt, H.-M., Schwefel, H.-P. (eds.) PPSN 1996. LNCS, vol. 1141, pp. 178–187. Springer, Heidelberg (1996)
23. Peierls, R.: On Ising's model of ferromagnetism. Proc. Cambridge Phil. Soc. 32, 477–481 (1936)
24. Santana, R.: Estimation of distribution algorithms with Kikuchi approximations. Evolutionary Computation 13(1), 67–97 (2005)
25. Yedidia, J.S., Freeman, W.T., Weiss, Y.: Constructing free energy approximations and generalized belief propagation algorithms. IEEE Transactions on Information Theory 51(7), 2282–2312 (2005)

Part III
Application

Chapter 12
Applications of Distribution Estimation Using Markov Network Modelling (DEUM)

John McCall, Alexander Brownlee, and Siddhartha Shakya

Abstract. In recent years, Markov Network EDAs have begun to find application to a range of important scientific and industrial problems. In this chapter we focus on several applications of Markov Network EDAs classified under the DEUM framework which estimates the overall distribution of fitness from a bitstring population. In Section 1 we briefly review the main features of the DEUM framework and highlight the principal features that have motivated the selection of applications. Sections 2 - 5 describe four separate applications: chemotherapy optimisation; dynamic pricing; agricultural biocontrol; and case-based feature selection. In Section 6 we summarise the lessons learned from these applications. These include: comparisons with other techniques such as GA and Bayesian Network EDAs; trade-offs between modelling cost and reduction in search effort; and the use of MN models for surrogate evaluation. We also present guidelines for further applications and future research.

12.1 Distribution Estimation Using Markov Network Modelling

Distribution Estimation Using Markov Network Modelling (DEUM) is an algorithmic framework for a family of Markov Network EDAs. DEUM algorithms operate on spaces of bitstrings and construct probabilistic models which estimate the

John McCall
Robert Gordon University, Aberdeen
e-mail: `j.mccall@rgu.ac.uk`

Alexander Brownlee
Loughborough University, Loughborough
e-mail: `A.E.I.Brownlee@lboro.ac.uk`

Siddhartha Shakya
Business Modelling and Operational Transformation Practice,
BT Innovate & Design, Ipswich
e-mail: `sid.shakya@bt.com`

S. Shakya and R. Santana (Eds.): Markov Networks in Evolutionary Computation, ALO 14, pp. 193–207.
springerlink.com

distribution of fitness over the search space as a Gibbs Distribution over the potential functions of the cliques of the network. Early publications on DEUM, including [26, 29], describe the relationship between the Gibbs energy $U(x)$ of a solution and its fitness $f(x)$. This is parameterised, and completely determined, by a set of Markov network parameters, one for each clique present in the network structure. Given sufficiently many evaluated solutions, Least Squares Approximation can be used to estimate these parameters and hence the distribution. The estimation has been observed to work best with a slightly over-specified system. A rigorous analysis of the population size needed in relation to a given structure is provided in Brownlee's thesis [3].

The precise form of the Gibbs Distribution is given in equation (12.1):

$$p(x) = \frac{e^{-U(x)/T}}{Z} \tag{12.1}$$

where,

$$Z = \sum_{y \in \Omega} e^{-U(y)/T} \tag{12.2}$$

Here, the numerator, $e^{-U(x)/T}$, represents the likelihood of a particular configuration x of the variables. The denominator, Z, is the normalising constant computed by summing over the set Ω of all possible configurations (note Z is never computed in practice). $U(x)$ is an *energy function* computed by summing potential contributions calculated from the values that x takes on each clique. Thus this exponentiated sum gives a factorisation of the distribution based on the structure of the Markov network. T is a temperature constant which controls the ruggedness of the probability distribution.

The key idea of the DEUM EDA is to model solution fitness as a mass distribution that equates to the Gibbs distribution as shown in equation (12.3):

$$p(x) = \frac{f(x)}{\sum_{y \in \Omega} f(y)} = \frac{e^{-U(x)/T}}{Z} \tag{12.3}$$

Thus sampling of $p(x)$ results in solutions of high fitness with a high probability.

More precisely, for any bitstring solution $x = (x_1, x_2, ..., x_n)$ we derive an equation (12.4) :

$$-ln(f(x)) = U(x) = \sum_{\kappa} \alpha_{\kappa} V_{\kappa}(x) \tag{12.4}$$

Here the sum is over all cliques κ. Each $V_{\kappa}(x)$ is a *clique potential function* which evaluates to ± 1 depending on the variable values x takes on clique κ. Once the parameters α have been estimated, negative exponentiation of $U(x)$ yields an estimate for $f(x)$. Thus equation (12.4) can be used as a model of the fitness function which we call the *Markov Fitness Model* (MFM).

The earliest, and most widely applied DEUM algorithm is DEUM$_d$. This assumes that there are no interactions between variables and so there is no structure. in this (univariate) case the MFM has the simple form given in equation (12.5):

$$-\ln(f(x)) = \alpha_0 + \alpha_1 x_1 + \alpha_2 x_2 + \ldots + \alpha_n x_n \quad (12.5)$$

Note that in equation (12.5) we abuse notation and use the shorthand x_i in place of $V_{\{x_i\}}$ by x_i. This evaluates to $+1$ if the variable value $x_i = 1$ and to -1 if $x_i = 0$. This shorthand carries over to larger cliques where the clique potential functions are interpreted, and written as the product of singleton cliques. Full details can be found in [3].

Other bivariate and multivariate forms of DEUM exist [3, 28] and have been shown to be competitive with other EDAS on important problems such as 2D-Ising Spin Glass and 3CNF-MAXSAT problems. The general form of a DEUM algorithm is as follows:

DEUM

---

1. Input or learn structure G.
2. Generate a population, P, consisting of M solutions
3. Select a set D from P consisting of N best solutions, where $N \leq M$.
4. For each solution, x, in D, build an equation of the form $-\ln(f(x)) = \sum_\kappa \alpha_\kappa V_\kappa(x)$ (where the sum is over all cliques $\kappa \subseteq G$)
5. Apply Least Squares to the system of N equations to estimate the parameter set $\{\alpha_\kappa\}$
6. Use $\{\alpha_\kappa\}$ to estimate the distribution $p(x)$
7. Apply Gibbs sampling to $p(x)$ to generate M new solutions to replace P and return to step 2 until termination criteria are met

---

Full technical details of DEUM can be found in [3, 28] and elsewhere in this book. The purpose of this chapter is to examine the performance of DEUM in relation to other evolutionary algorithms (EAs), and in particular other EDAs when applied to real-world problems.

12.2 Chemotherapy Optimisation

Cancer is a serious and often fatal disease, widespread in the developed world. One of the most common methods of treating cancer is chemotherapy with toxic drugs. These drugs themselves can often have debilitating or life-threatening effects on patients undergoing chemotherapy. Clinicians attempting to treat cancer are therefore faced with the complex problem of how to balance the need for tumor treatment against the undesirable and dangerous side-effects that the treatment may cause. In this chapter we will consider ways in which EAs can be used to assist clinicians in developing approaches to chemotherapy that offer improvements over current clinical practice.

12.2.1 Cancer and Chemotherapy

Cancer is a broad term covering a range of conditions known as neo-plastic disease [33]. All forms of cancer have in common the phenomenon of uncontrolled cell proliferation. The result is the build up over time of more and more unhealthy cells, ultimately leading to death. A cytotoxic drug (also termed a ÒcytotoxicÓ) is a chemical that destroys the cells it is exposed to. Chemotherapy is the use of cytotoxics to control or eliminate cancer. The drugs are administered to the body by a variety of routes, the objective of which is to create a concentration of drug in the bloodstream. A concentration of a cytotoxic in the bloodstream will act, therefore, to kill cells systemically. This means that both tumour cells and healthy cells in other organs of the body will alike be killed by the drugs. The intention is to eradicate the tumour or at least to control its growth. However the treatment has toxic side-effects on the rest of the body. The success of chemotherapy depends crucially on maintaining damage to the tumour whilst effectively managing the toxic side-effects.

12.2.2 Optimisation of Cancer Chemotherapy

A major goal of cancer medicine is the selection of a treatment that provides, for an individual patient, the best outcome as measured against a set of clinical objectives. There is now a considerable body of research that attempts to formulate and solve chemotherapy optimisation as an optimal control problem. Almost universally, these works use a differential equation-based simulation of the natural growth of the tumour and the loss in cells due to the action of cytotoxics. These tumour response models are non-linear and constrained and so are often intractable to traditional optimisation techniques. Hence there have been a number of applications of evolutionary algorithms and, more recently, estimation of distribution algorithms to chemotherapy optimisation. A detailed discussion of this literature can be found in [15].

Here we focus on the formulation by Martin *et al.* [14] of chemotherapy as a single-objective optimisation, where the aim is to select a sequence $C = \{C_{ij}\}$ of doses of cytotoxics to minimise the tumour burden, equation (12.6), subject to the state equation (12.7) and constraint equations (12.8 - 12.11).

$$J(C) = \int_0^T N(t)dt \tag{12.6}$$

$$\frac{dN}{dt} = N(t)[\lambda \ln \frac{\Theta}{N(t)}] - \sum_{j=1}^{d} \kappa_j \sum_{i=1}^{n} C_{ij}\{H(t-t_i) - H(t-t_{i+1})\} \tag{12.7}$$

$$C_{ij} \geq 0 \;\; \forall i = 1,...,n, j = 1,...,d \tag{12.8}$$

$$C_{cumj} \geq \sum_{i=1}^{n} C_{ij} \;\; \forall j = 1,...,d \tag{12.9}$$

$$N_{max} \geq N(t_i) \;\; \forall i = 1,...,n \tag{12.10}$$

$$C_{s-effk} \geq \sum_{j=1}^{d} \eta_{kj} C_{ij} \;\; \forall i = 1,...,n, k = 1,...,m \tag{12.11}$$

Here, $N(t)$ represents the number of cells in the tumour at time t, λ is the tumour growth rate and Θ is an assymptotic limiting size for the tumour under unrestrained growth. The coefficients κ_j control the potencies of each of d drugs in terms of proportionate cell-kill and H is the Heaviside step function. C_{cumj} is the maximum cumulative dose of drug j allowed during the treatment period. N_{max} is the maximum size to which the tumour may be allowed to grow during the treatment period. C_{s-effk} is an upper bound on the severity of side-effect k permissible at any point in the treatment period and the coefficients η_{kj} represent the contributions of drug j to toxic side-effect k. Side-effects modelled include nausea and vomiting, hair loss and damage to healthy tissues such as the heart, bone marrow and lungs.

12.2.3 EDAs for Cancer Chemotherapy Optimisation

Multi-drug chemotherapy schedules may be usefully represented by decision vectors $C = \{C_{ij}\}$ where $1 \leq i \leq n$, $1 \leq j \leq d$, encoded as binary strings with four bits each allocated to the drug dose C_{ij} (the ith dose of drug j) [25]. Thus each drug dose takes an integer value in the range of 0 to 15 concentration units. In general, with n treatment intervals and up to 2^p concentration levels for d drugs, there are up to 2^{npd} possible chemotherapies. Henceforth we assume that $n = 10$ and that $d = 10$ [23]. This gives a length 400 bitstring search space of 2^{400} candidate chemotherapies.

The fitness function is derived from equations (12.7 - 12.11) as follows. First, The continuous integral is discretised into a series of n time steps, over which the tumour size is assumed constant. Drug doses are assumed to have an instantaneous effect at the beginning of a time step, and the growth equation is similarly assumed to operate in a discrete manner. This is a reasonable approximation when viewed over a treatment period lasting several months [23]. Second a substitution $u(t) = \ln \frac{\Theta}{N(t)}$ is made. This has the effect of converting the tumour burden minimisation problem into a maximisation problem and also simplifies calculation [14]. Finally, the four types of constraints defined in equations (12.8 - 12.11) are incorporated as penalty terms proportional to the magnitudes of violation.

This results in the evaluation function given in equation (12.12) for each solution $C = \{C_{ij}\}$:

$$F(C) = \sum_{i=1}^{n} \sum_{j=1}^{d} \kappa_j \sum_{i=1}^{n} C_{ij} e^{\lambda(t_{i-1} - t_n)} - \sum_{s=1}^{4} P_s D_s \tag{12.12}$$

In [26], [15] and [4] the EDAs PBIL, DEUM$_d$, UMDA and hBOA are variously applied to the chemotherapy optimisation problem using the encoding and penalised objective function given above. In each of these works, algorithms are assessed on efficiency and solution quality. Efficiency is measured by the number of solution evaluations required on average to find a feasible solution. Solution quality is measured by the value of F obtained after a fixed number of solution evaluations, much greater than that required for feasibility. In [26], it was shown that the PBIL significantly outperforms a GA on both efficiency and solution quality. In [15], it is shown that DEUM$_d$ significantly outperforms both PBIL and a GA on this problem, in terms of both efficiency and solution quality. However the improvement over PBIL is much less marked than the advantage both EDAs have over GAs. Finally, in [4], DEUM$_d$ is compared with another univariate EDA, UMDA [16] and the Bayesian network-based multivariate EDA hBOA [22]. Here hBOA detects a large number of interactions between problem variables indicating that this is a multivariate problem. However, somewhat surprisingly, it underperforms the univariate EDAs on both efficiency and quality. The authors conclude that the multivariate interactions are *unnecessary* and in fact impede the search by hBOA. In this last work, UMDA is significantly more efficient than DEUM$_d$ and PBIL, though DEUM$_d$ remains the best overall on solution quality.

Chemotherapy optimisation is complex, non-linear, and highly-constrained with hundreds of variables. DEUM$_d$ has proved competitive with leading EDAs and other EAs on this problem, in many cases proving significantly better on both efficiency and solution quality. It is believed that the precision of the MFM has proved advantageous in this problem which has a moderate degree of multivariate interaction. Finally, it is worth noting that, in simulation at least, all algorithms produce solutions that outperform standard chemotherapies used in clinical practice [24].

12.3 Dynamic Pricing

E-commerce has transformed the way firms price their products and interact with their customers. The increasingly dynamic nature of the e-commerce has produced a shift away from fixed pricing to dynamic pricing [2]. The basic idea of dynamic pricing (also called flexible pricing [17]) is for a firm to adjust the price of its products or services, online, as a function of its perceived demand at different times for the different price levels. Traditionally, dynamic pricing strategies have been applied in service industries. More recently, dynamic pricing is also being used in other non-traditional domains, such as, supply chain management, and planning and product manufacture. A significant increase in profit has been reported by companies implementing dynamic pricing strategies, resulting in increasing interest in this area.

In this section, we describe recent work that formulates profit maximization through dynamic pricing as a metaheuristic search and investigates the relative performances of EAs, and in particular two EDAs including DEUM$_d$ on this problem [30], [31].

12.3.1 Dynamic Pricing for Resource Management

This work was motivated by an automated resource management system, [20][32], which manages access services and provides telecommunication service to its customers. Resource management can be loosely defined as the effective workforce utilization for a given calendarised work demand profile, while satisfying constraints such as quality of service, overtime and borrowing additional workforce [19]. The work presented in this section focuses on investigating the use of dynamic pricing for improving resource management. The aim is to increase profit by optimizing the use of resources at any given time period during the life cycle of a product (or service). In this subsection we describe a dynamic pricing model that can be used for analyzing both short-term (weeks or days) and long-term (months or years) profit from a product.

In [30], [31], the authors derive the following model, expressed in equation (12.13) for the total profit made over a period assuming dynamic pricing is used to manage demand:

$$\Pi = \sum_{t=1}^{N} \left[a_{0t} Q_t + \sum_{j=1}^{N} a_{jt} Q_j Q_t - C_t Q_t \right] (1+\varepsilon) \tag{12.13}$$

Here, Π represents the total profit earned during the entire planning horizon, N represents the number of planning periods, t denotes any particular planning period, Q_t is the managed demand (thought of as the number of jobs produced) during period t, P_t is the average price of a job set in period t, and C_t represents the cost of supplying one extra job during period t. The coefficients a_{ij} relate price P_i in period i to demand over all periods j according to the following equation (12.14):

$$P_t = a_{0t} + a_{1t} Q_1 + a_{2t} Q_2 + \ldots + a_{tt} Q_t + \ldots + a_{Nt} Q_N \tag{12.14}$$

In normal circumstances, $a_{tt} < 0$, that is an increase in number of jobs available in a given period leads to a decrease in the price which can be charged in that period. The stochastic shock component ε represents random error in profit due to random fluctuations in demand and costs and is drawn from a normal distribution, $\varepsilon \approx N(0, \sigma)$ with standard deviation σ. We can control σ to simulate varying levels of stochasticity in the dynamic market.

Finally, some constraints are imposed to reflect upper and lower bounds on resource capacity.

a. *Available capacity constraints* - These are number of jobs that can be produced in a given period and regulates the resources such as number of workers or machines that should be used in a given period. For all $t = 1,...,N$
$M_t \leq Q_t$ - Lower bound for the capacity constraint
$K_t \geq Q_t$ - Upper bound for the capacity constraint
b. *Price capacity constraints* - These are the prices of a job (product) produced in a given period and regulate the value to the costumers. For all $t = 1,...,N$
$\overline{P}_t \leq P_t$ - Lower bound for the capacity constraint
$\underline{P}_t \geq P_t$ - Upper bound for the capacity constraint
Equations (12.13), (12.14) and the associated capacity constraints define a constrained, non-linear, stochastic optimisation problem. We now consider application of EA to this problem.

12.3.2 Evolutionary Algorithms for Resource Management Using a Dynamic Pricing Model

A solution, x, is represented as a set of managed demands $Q = \{Q_1, Q_2, ..., Q_N\}$, where each Q_t is represented by a bit-string of length l. The total length of a bit-string solution, $x = x_1, x_2, ..., x_n$, is therefore, equal to $n = l \times N$. The range for each set of l bits representing Q_t is set from M_t to K_t, enforcing the capacity constraints. Profit can now be considered as a stochastic function $\Pi = \Pi(x)$. Adopting an approach from [21], the authors in [30], [31] incorporate price constraints by defining a penalised objective function as:

$$F(x) = \Pi(x) - h(k)H(x) \tag{12.15}$$

where $h(k)$ is a dynamically modified penalty value, k is the algorithm's current iteration number, and $H(x)$ is a penalty factor, based on the absolute magnitudes of violations of the pricing constraints. Full details are available in [30], [31] .

Different problem scenarios are set up by varying the price-demand coefficients a_{ij} and the level, σ, of stochastic shock in demand. For short-term analysis, periods are taken to be days and the a_{ij} are set so that the production for a given day is a negative function of the price on that day and a positive function of the prices on other days of the week. For long-term analysis, periods are taken to be years and the a_{ij} are set so that the production (demand) in a year is a negative function of the average price in that year and positive function of the production during the previous year. For both short and long term analysis, σ is varied over the set $\{0.0, 0.1, ..., 1.0\}$. This represents varying levels of stochastic shock in the actual demand encountered during each period.

Two EDAs and a GA were applied to this problem. They were PBIL [1], DEUM$_d$, and a GA [11]. In addition, the authors also applied Simulated Annealing (SA) [13] to this problem.

PBIL and DEUM$_d$, are both univariate EDAs and so assume problem variables are mutually independent. The motivation for using univariate EDAs for this problem is two fold: firstly, they are simple and, therefore, often quickly converge to optima resulting in higher efficiency. This is particularly important since efficiency of an algorithm matters a lot in a dynamic environment. Secondly, the number of problems that has been shown to be solved by univariate EDAs is surprisingly large.

In [31], a total of 100 executions of each algorithm was done for a range of short and long term scenarios and the best policy together with the total profit found in each execution was recorded. Also, a reliability factor (*RL*) was calculated, that is the total percentage of runs where the final population of the algorithm converged to a policy satisfying all the constraints. A high *RL* is desirable since it indicates a high probability of achieving at least the suggested profit with the proposed solution. It was demonstrated experimentally that SA had the worst performance in terms of profit maximisation and also the lowest *RL*. The performances of DEUM_d and PBIL were broadly comparable. In terms of mean profit, PBIL and DEUM_d had the best performances of all algorithms. However, in the presence of uncertainty in demand, GA has the highest *RL* and therefore has the overall best performance. It was observed that, in general, demand uncertainty reduces the total profit, and is true for all the tested algorithms. Full results for all dynamic pricing scenarios tested with these algorithms can be found in [30] and [31] .

We conclude that DEUM_d is a competitive algorithm for the resource management through dynamic pricing application and make an additional observation regarding the encoding used here. All of the experiments assume a planning horizon of seven periods, with the maximum demand in each period represented by a length 12 bitstring. This results in an overall bitstring solution of length 84. DEUM_d therefore requires $7 \times 12 = 84$ model coefficients to be determined on each population. The authors used populations of size 10 in each experiment, so the models were underspecified. Work in [3] suggests a population size of around 90 would have produced the best possible models. However this would have caused an increase in calculation time. Also this would have reduced the number of iterations since total evaluations was fixed. In this case, the Singular Value Decomposition algorithm used to determine model parameters [28] will have balanced unmodelled variation across parameters. In this case, that proved adequate to the problem.

12.4 Agricultural Bio-control

When mushrooms are produced in commercial quantities, the quality and yield of the mushroom crop can be seriously damaged through infestation by sciarid flies. Sciarid fly larvae are known to feed on the mycelium in the casing layer of mushroom causing crop production to significantly decline. An important weapon in combatting sciarid fly is the use of the nematode worm, *Steinernema feltiae*, which feeds on sciarid larvae thus reducing the problem. Nematode worms are sold commercially for bio-control of sciarid flies on mushroom farms.

In [9], a dynamic mathematical model is developed that expresses the life cycle of Sciarid larvae in the presence of periodic spraying with nematode worms. In this model, the mushrooms are sprayed with nematodes at a series of discrete interventions points. The model can be implemented numerically using standard techniques and consists of a set of coupled differential equations representing the 4 stages of the sciarid life cycle (egg, larva, pupa and adult) and the nematode worm. There are transitions between sciarid populations representing the growth stages of the flies and there are interaction terms representing fatal interactions with nematode worms. Sciarids are damaging to the mushroom crop only in the larval stage (though adults lay eggs which can then develop into larvae) and so the main focus of bio-control is to minimise the number of larvae at any time. Fenton et. al. [9] developed optimisation objectives based on the number, $L(t)$, of larvae present in the crop at time t. The problem admits a *bang-bang control* strategy, that is a fixed dose of nematodes is either applied or not applied at each of a series of potential intervention points during the period of mushroom cultivation. Spraying the crop with nematodes is both costly and time consuming and so there is a second objective that aims to spray efficiently, i.e. to minimise the number of sprayings required. This is clearly in conflict with the objective of reducing the larva population. From [9], [10], a weighted objective function has been formulated as follows:

$$F(x) = \sum_{t=0}^{T} L(t) + N(x)P \tag{12.16}$$

Here x is a bitstring representing a bang-bang control strategy, encoding a series of binary intervention / non-intervention decisions at a series of evenly-spaced time steps $t = 0, 1, ..., T$. A control strategy evaluates as $F(x)$ which sums the number of larvae present at each time step and finally adds on a penalty P for each of $N(x)$ interventions specified by x. The values of $L(t)$ are calculated by running the bio-control model with the specified control and recording the size of the larval population at each time step.

12.4.1 Evolutionary Algorithms for Agricultural Bio-control

In [35], DEUM$_d$ was applied to the mushroom bio-control problem and performance was compared against a standard GA and two GAs incorporating *directed crossover* operators, TInSSel and CalEB [10]. A further paper [5] also applied a bivariate algorithm, Chain DEUM to the problem. This latter introduced structure relating successive time intervals, corresponding to switching times in the bang-bang control. The MFM for Chain DEUM can be written as in equation (12.17)

$$-\ln(F(x)) = \alpha_0 + \sum_{i=1}^{n} \alpha_i x_i + \sum_{j=1}^{n-1} \beta_j x_j x_{j+1} \tag{12.17}$$

where the coefficients β_j are the MFM parameters associated to those cliques relating the variables for consecutive decision periods. The results showed DEUM$_d$ to be competitive with or better than the GAs on both reducing larvae and minimising the number of doses. Chain DEUM outperformed all of the other algorithms. Statistics from repeated runs showed that Chain DEUM distributed sprayings consistently and accurately to coincide with the natural breeding cycle of the larvae, thus maximising the effectiveness of the nematodes. The other algorithms, particularly the GAs, were less consistent and so would be less efficient in determining optimal spraying times. It was further observed that building the Chain DEUM MFM in equation (12.17) resulted in parameters that could be used directly to infer near optimal dosing points without iterative optimisation, though with less accuracy. In this application therefore, the Markov network approach has proved highly effective in terms of optimisation and produced model parameters that have meaningful interpretation in terms of the application domain.

12.5 Feature Selection in Case-Based Reasoning

Case-based reasoning (CBR) solves new problems based on the solutions of similar past problems. Cases describing the problem faced and the solution applied are stored in a case base for future retrieval when a similar new problem is encountered. CBR is often applied to classification tasks (e.g. for decision support or diagnostic problems) in which the problem is represented as a feature vector and the solution by a class label. In classification tasks a good feature is one that is predictive of the problem class on its own or in combination with other features. The problem is that not all features are important: some may be redundant; some may be irrelevant; and some can even be harmful to further analysis. Instance-based learners are very susceptible to irrelevant features and applying a more compact representation has been shown to successfully improve accuracy, as in [12]. In addition, reducing the number of features increases speed of analysis and alleviates the problem that some learning algorithms break down with high dimensional data.

Approaches to feature selection employed by the CBR community can be categorised into filter and wrapper methods. Filters are data pre-processors that do not require feedback from the final learner, such as information gain or the Chi-squared score described by [36]. As a result they tend to be faster, scaling better to large datasets; see [34]. The wrapper approach uses feedback from the final learning algorithm to guide the search for a subset of features; examples include forward selection, backward elimination and GAs as in [8]. A recent, comprehensive review of feature subset selection methods can be found in [27]. Generally, feedback ensures wrappers select a better set of features tailored for the learning algorithm corresponding to a higher fitness for the GA. Unfortunately it has the disadvantage of being time consuming because feedback involves learner accuracy ascertained from cross-validation runs. On larger datasets the time taken by wrapper approaches tends to offset the improvement in results; our work aims to reduce the run time while still maintaining the improved results.

12.5.1 Genetic Algorithm Using MFM Surrogate

Solutions to the feature selection problem may be encoded as bitstrings where each bit represents a feature. The feature is selected for classification if the bit is set to 1. In the other case, the feature is not selected. Solutions are evaluated by the following (expensive) process. First the fitness function takes feedback from a leave-one-out accuracy evaluation with a 3-Nearest Neighbour classifier. A filter approach is then applied to optimise the feature subset selection for the generated model. Finally a floating point value is returned representing the accuracy of the classifier with the selection of features represented by the solution.

In [7], instead of a direct application of a Markov-network EDA, the authors use the MFM model (12.4) as a surrogate for the fitness function in a genetic algorithm. The model used for the MFM combines terms representing bivariate structure learned by the algorithm described below and the full set of univariate terms. The aim is to gain an efficient algorithm that can be applied to larger datasets than traditional wrapper methods; and selection of good feature subsets that give higher accuracies than filter methods. The algorithm, MFM-GA, is defined as follows:

MFM-GA

---

1. Initialise randomly-generated population p_1
2. Evaluate fitnesses of p_1 using the true fitness function
3. Select a subset σ_1, the top 25% of the population p_1
4. Run Chi-square structure learning algorithm to find statistical dependencies apparent in σ_1
5. Initialise randomly-generated population p_2
6. Evaluate fitnesses of p_2 using true fitness function
7. Select a subset σ_2 of the population p_2
8. Compute model parameters for MFM from σ_2
9. Run standard genetic algorithm:
10. **repeat**
    a. Evaluate p_3 using Markov network surrogate
    b. Select a subset σ_3 of p_3
    c. Generate new population, replacing p_3 using standard crossover and mutation operators on σ_3, preserving elite individuals
11. **until** surrogate evaluation limit reached

The algorithm runs in three distinct phases. Steps 1 to 4 are the structure learning phase, where the structure to be used for the Markov network is learned from a sample population. In steps 5 to 8, the parameters of the MFM are estimated from the sample population. At this point we have now constructed the surrogate fitness model and we can now proceed to run the GA. This takes place in steps 9 to 11; during this phase no evaluations of the true fitness function are made and the surrogate is used to assign fitness values to individuals in its place.

Experiments were conducted using two datasets, Sonar and Vehicle from the UCI collection. Relationships between variables of these datasets are not defined and therefore, the exact structure of the problem is not known. Consequently, a Markov network structure learning algorithm using Chi-Square independence tests [6] was used in Steps 1 to 4. Using a threshold of 10, the Chi-Square algorithm discovered 16 and 7 bivariate interactions in the Sonar and Vehicle datasets respectively. These were added to the full set of univariate terms and then set as the structure of the MFM before running the GA component of the algorithm. The MFM-GA was compared with the same GA using only true evaluations. Significant speed-ups were noted at the penalty of a small reduction in solution quality, most notably for the Vehicle dataset. MFM-GA spent most of its time in structure learning. When the structure was supplied (as might be possible using expert knowledge) speed-ups were particularly marked. In all cases MFM-GA outperformed the traditional information gain method on solution quality.

12.6 Conclusion

The DEUM group of Markov network EDAs are applicable to a wide range of nonlinear, constrained discrete optimisation problems. Here we have examined applications to cancer chemotherapy, resource management using dynamic pricing, agricultural biocontrol and feature selection for case based reasoning. DEUM has proved competitive with other EDAs on these problems. Advantages of the approach include: explicit modelling of fitness distribution so that high fitness solutions are sampled with high probability; the discovery of Markov network parameters which can contain useful problem knowledge; and the potential uses of the MFM as a fitness surrogate. Disadvantages include the computational overhead of both modelling and sampling and the current restriction to bitstring encodings. Applications of the DEUM approach are likely to be most beneficial when the fitness function is sufficiently expensive that the reduction in fitness evaluations gained from better modelling outweighs the modelling and sampling costs. Future research, as for other EDA approaches, is likely to focus on cheaper modelling, incremental model updates and the hybridisation of MFM estimation with other algorithms.

References

1. Baluja, S.: Population-based incremental learning: A method for integrating genetic search based function optimization and competitive learning. Technical Report CMU-CS-94-163, Pittsburgh, PA (1994)
2. Bichler, M., Kalagnanam, J., Katircioglu, K., King, A.J., Lawrence, R.D., Lee, H.S., Lin, G.Y., Lu, Y.: Applications of flexible pricing in business-to-business electronic commerce. IBM Systems Journal 41(2), 287–302 (2002)
3. Brownlee, A.E.I.: Multivariate Markov Networks for Fitness Modelling in an Estimation of Distribution Algorithm. PhD thesis, The Robert Gordon University, Aberdeen, UK (May 2009)
4. Brownlee, A., Pelikan, M., McCall, J., Petrovski, A.: An Application of a Multivariate Estimation of Distribution Algorithm to Cancer Chemotherapy. In: Proc. ACM GECCO 2008, pp. 463–464 (2008)

5. Brownlee, A., Wu, Y., McCall, J., Godley, P., Cairns, D., Cowie, J.: Optimisation and Fitness Modelling of Bio-control in Mushroom Farming using a Markov Network EDA. In: Proc. ACM GECCO 2008, pp. 465–466 (2008)
6. Brownlee, A., McCall, J., Shakya, S., Zhang, Q.: Structure Learning and Optimisation in a Markov-network based Estimation of Distribution Algorithm. In: Proc. IEEE CEC 2009, pp. 447–454 (2009)
7. Brownlee, A., Regnier-Coudert, O., McCall, J., Massie, S.: Using a Markov network as a surrogate fitness function in a genetic algorithm. In: Proc. IEEE CEC 2010, pp. 4525–4532 (2010)
8. Das, S.: Filters, Wrappers and a Boosting-Based Hybrid for Feature Selection. In: Proc. ICML 2001, pp. 74–81. Morgan Kaufmann Publishers Inc. (2001)
9. Fenton, A., Gwynn, R.L., Gupta, A., Norman, R., Fairbairn, J.P., Hudson, P.J.: Optimal application strategies for entomopathogenic nematodes: integrating theoretical and empirical approaches. Journal of Applied Ecology 39, 481–492 (2002)
10. Godley, P.M., Cairns, D.E., Cowie, J.: Directed intervention crossover applied to bio-control scheduling. In: Proc. IEEE CEC 2007, pp. 638–645 (2007)
11. Goldberg, D.: Genetic Algorithms in Search, Optimization, and Machine Learning. Addison-Wesley (1989)
12. John, G.H., Kohavi, R., Pfleger, K.: Irrelevant features and the subset selection problem. In: Proc. ICML 1994, pp. 121–129. Morgan Kaufmann Publishers Inc. (1994)
13. Kirkpatrick, S., Gelatt, C.D., Vecchi, M.P.: Optimization by simulated annealing. Science 220(4598), 671–680 (1983)
14. Martin, R.B., Teo, K.L.: Optimal Control of Drug Administration in Cancer Chemotherapy. World Scientific, Singapore (1994)
15. McCall, J., Petrovski, A., Shakya, S.: Evolutionary Algorithms for Cancer Chemotherapy Optimization. In: Fogel, G., Corne, D., Pan, Y. (eds.) Computational Intelligence in Bioinformatics, ch. 12, pp. 265–296. Wiley Interscience (2008)
16. Mühlenbein, H., Paaß, G.: From Recombination of Genes to the Estimation of Distributions I. Binary Parameters. In: Ebeling, W., Rechenberg, I., Voigt, H.-M., Schwefel, H.-P. (eds.) PPSN 1996. LNCS, vol. 1141, pp. 178–187. Springer, Heidelberg (1996)
17. Narahari, Y., Raju, C.V., Ravikumar, K., Shah, S.: Dynamic pricing models for electronic business. Sadhana 30(part 2,3), 231–256 (2005)
18. Oliveira, F.S.: A constraint logic programming algorithm for modeling dynamic pricing. Informs Journal of Computing 30, 69–77 (2007)
19. Owusu, G., Dorne, R., Voudouris, C., Lesaint, D.: Dynamic planner: A decision support tool for resource planning, applications and innovations in intelligent systems. In: Proc. of ES 2002, pp. 19–31 (2002)
20. Owusu, G., Voudouris, C., Kern, M., Garyfalos, A., Anim-Ansah, G., Virginas, B.: On optimising resource planning in BT with FOS. In: Proc. International Conference on Service Systems and Service Management, pp. 541–546 (2006)
21. Parsopoulos, K., Vrahatis, M.: Particle swarm optimization method for constrained optimization problems. In: Intelligent Technologies - Theory and Application: New Trends in Intelligent Technologies. Frontiers in Artificial Intelligence and Applications, vol. 76, pp. 214–220 (2002)
22. Pelikan, M.: Hierarchical Bayesian Optimization Algorithms. Springer (2005)
23. Petrovski, A.: An Application of Genetic Algorithms to Chemotherapy Treatment. PhD Thesis, The Robert Gordon University, Aberdeen (1999)

24. Petrovski, A., McCall, J.: Computational Optimization of Cancer Chemotherapies using Genetic Algorithms. In: John, R., Birkenhead, R. (eds.) Soft Computing Techniques and Applications, pp. 117–122. Physica-Verlag, Heidelberg (2000)
25. Petrovski, A., McCall, J.A.W.: Multi-objective Optimisation of Cancer Chemotherapy Using Evolutionary Algorithms. In: Zitzler, E., Deb, K., Thiele, L., Coello Coello, C.A., Corne, D.W. (eds.) EMO 2001. LNCS, vol. 1993, pp. 531–545. Springer, Heidelberg (2001)
26. Petrovski, A., Shakya, S., McCall, J.: Optimising cancer chemotherapy using an estimation of distribution algorithm and genetic algorithms. In: Proc. ACM GECCO 2006, pp. 413–418 (2006)
27. Saeys, Y., Inza, I., Larrañaga, P.: A review of feature selection techniques in bioinformatics. Bioinformatics 23(19), 2507–2517 (2007)
28. Shakya, S.: DEUM: A Framework for an Estimation of Distribution Algorithm based on Markov Random Fields. PhD thesis, The Robert Gordon University, Aberdeen, UK (April 2006)
29. Shakya, S., McCall, J., Brown, D.: Using a Markov Network Model in a Univariate EDA: An Empirical Cost-Benefit Analysis. In: Proc. ACM GECCO, pp. 727–734 (2005)
30. Shakya, S., Oliveira, F., Owusu, G.: An application of EDA and GA to dynamic pricing. In: Proc. ACM GECCO 2007, pp. 585–592 (2007)
31. Shakya, S., Oliveira, F., Owusu, G.: Analysing the Effect of Demand Uncertainty in Dynamic Pricing with EAs. In: Bramer, M., Coenen, F., Petridis, M. (eds.) Proc. AI 2008, pp. 77–90. Springer (2009)
32. Voudouris, C., Owusu, G., Dorne, R., McCormick, A.: FOS: An advanced planning and scheduling suite for service operations. In: Proc. International Conference on Service Systems and Service Management, pp. 1138–1143 (2006)
33. Wheldon, T.E.: Mathematical Models in Cancer Research. Adam Hilger, Bristol (1988)
34. Wiratunga, N., Koychev, I., Massie, S.: Feature Selection and Generalisation for Retrieval of Textual Cases. In: Funk, P., González Calero, P.A. (eds.) ECCBR 2004. LNCS (LNAI), vol. 3155, pp. 806–820. Springer, Heidelberg (2004)
35. Wu, Y., McCall, J., Godley, P., Brownlee, A., Cairns, D., Cowie, J.: Bio-control in mushroom farming using a Markov Network EDA. In: Proc. 2008 IEEE CEC 2008, pp. 2996–3001 (2008)
36. Yang, Y., Pedersen, J.: A Comparative Study on Feature Selection in Text Categorization. In: Proc. ICML 1997, pp. 412–420. Morgan Kaufmann Publishers Inc. (1997)

Chapter 13
Vine Estimation of Distribution Algorithms with Application to Molecular Docking

Marta Soto, Alberto Ochoa, Yasser González-Fernández, Yanely Milanés, Adriel Álvarez, Diana Carrera, and Ernesto Moreno

Abstract. Four undirected graphical models based on copula theory are investigated in relation to their use within an estimation of distribution algorithm (EDA) to address the molecular docking problem. The simplest algorithms considered are built on top of the product and normal copulas. The other two construct high-dimensional dependence models using the powerful and flexible concept of vine-copula. Empirical investigation with a set of molecular complexes used as test systems shows state-of-the-art performance of the copula-based EDAs in the docking problem. The results also show that the vine-based algorithms are more efficient, robust and flexible than the other two. This might suggest that the use of vines opens new research opportunities to more appropriate modeling of search distributions in evolutionary optimization.

13.1 Introduction

Estimation of distribution algorithms (EDAs) [25, 28] have been used successfully in several Bioinformatic problems [3], such as *de novo* peptide design [7], protein folding [33] and protein side chain placement [34]. In this chapter we investigate the performance of several EDAs based on undirected copula models in the molecular docking problem.

Marta Soto · Alberto Ochoa · Yasser González-Fernández
Institute of Cybernetics, Mathematics, and Physics, Cuba
e-mail: {mrosa,ochoa,ygf}@icimaf.cu

Yanely Milanés · Adriel Álvarez · Diana Carrera
University of Havana, Cuba
e-mail: {y.milanes,a.mosquera,d.carrera}@lab.matcom.uh.cu

Ernesto Moreno
Center of Molecular Immunology, Cuba
e-mail: emoreno@cim.sld.cu

S. Shakya and R. Santana (Eds.): Markov Networks in Evolutionary Computation, ALO 14, pp. 209–225.
springerlink.com

Molecular docking consists in predicting the binding geometry of small flexible ligands that bind a large macromolecular entity, in most cases a protein. Currently, docking has become an important component of structure-based drug design, with the aim of optimizing the initial screening phases of the long and costly process for developing a new drug.

In this chapter, several undirected graphical models based on copula theory are combined into four different estimation of distribution algorithms to address the molecular docking problem. The algorithms are based on product, normal copulas and vines. The later model is a powerful and flexible class of high-dimensional dependency models, which represent a rich variety of dependencies and marginal distributions. We pursue two goals: to test the potential of EDAs in the docking problem and evaluate the relative performance of EDAs based on vines in comparison to unstructured copula models.

The outline of the chapter is as follows. Section 13.2 gives a necessary background on copula theory and describes two EDAs, UMDA and GCEDA, which are based on the product and normal copulas, respectively. Section 13.3 describes two algorithms based on the copula-vine models, DVEDA and CVEDA. Section 13.4 provides a brief introduction on the molecular docking problem and the fitness function model. The empirical investigation is reported in Section 13.5. Numerical results with the genetic, particle swarm optimization and differential evolution algorithms are included in Section 13.6. The conclusions are given in Section 13.7.

13.2 Two Continuous EDAs Based on Undirected Models

Nowadays, there is an increasing interest on EDA approaches based on copula theory. Several copula-based EDAs have been proposed in the literature, for example, [12, 39, 40]. In this section we briefly describe two multivariate copula-based EDAs that are relevant to this work.

A multivariate normal distribution forms a Markov random field (MRF) with respect to a graph $G = (V,E)$ if the missing edges correspond to zeros on the precision matrix (the inverse covariance matrix) $X = (X_i)_{i \in V} \sim N(\mu,\Sigma)$, so that $\left(\Sigma^{-1}\right)_{ij} = 0$ if $\{i,j\} \notin E$.

As important subclasses of this Gaussian MRF one can list the models associated with the independence graph, trees and join trees, and also several models based on copula theory. In this chapter we focus on the later class of algorithms. We start with some definitions from the copula theory.

A comprehensive treatment on copula theory is given in [21, 30]. The notion of copulas separates the effect of dependence from the effect of margins (scale, localization) in a joint distribution [24]. In this sense, copulas are functions that join multivariate distributions to their margins. The theory supports this definition with the Sklar's theorem [35].

Let $\mathbf{X} = (X_1,\ldots,X_n)$ and $\mathbf{x} = (x_1,\ldots,x_n)$ denote a continuous random vector and one of its possible configurations, respectively. Let also f, F and $F_1,\ldots,F_n$ be the joint density function, the joint cumulative distribution function and the margins associated to **X**. The basic relations supported by the Sklar's theorem are

$$F(x_1,\ldots,x_n) = C(F(x_1),\ldots,F(x_n))$$

and

$$C(u_1,\ldots,u_n) = F\left(F_1^{(-1)}(u_1),\ldots,F_n^{(-1)}(u_n)\right),$$

where C is a unique cumulative distribution function with uniform marginal distributions in $[0,1]$.

An immediate consequence of Sklar's theorem is that the random variables in a vector are independent if and only if their underlying copula corresponds to the product copula, which is given by

$$C_I(u_1,\ldots,u_n) = u_1.\ldots.u_n. \qquad (13.1)$$

The famous Univariate Marginal Distribution Algorithm (UMDA) [25] for continuous variables is obviously connected to (13.1) because it entails independence among all variables.

Another important copula is the multivariate normal copula,

$$C_N(u_1,\ldots,u_n;R) = \Phi_R\left(\Phi^{-1}(u_1),\ldots,\Phi^{-1}(u_n)\right), \qquad (13.2)$$

where Φ_R is the standard multivariate normal distribution function with linear correlation matrix R and Φ^{-1} is the inverse of the standard univariate normal distribution. Although in (13.2), Φ is the standard normal distribution, this copula can also be associated with other types of marginal distributions. If that is the case, the created joint density is not longer the multivariate normal, although the normal dependence structure is preserved.

With non-normal margins, the correlation matrix is estimated using either $\hat{R}_{ij} = sin\left(\pi/2\hat{\tau}_{ij}\right)$ or $\hat{R}_{ij} = 2sin\left(\pi/6\hat{\rho}_{s_{ij}}\right)$ $(i,j = 1,\ldots,n$ for each pair of variables $i,j)$. Here $\hat{\tau}$ and $\hat{\rho}_s$ are the non-parametric estimators Kendall's tau and Spearman's rho, respectively. If the resulting $\hat{R}$ is not positive-definite the correction proposed in [32] is applied.

Our second algorithm, the Gaussian Copula EDA (GCEDA) [37], was designed based on the multivariate normal copula.

As it was pointed out, a key advantage of the models based on copula is their ability to deal with different types of marginal distributions. In this work, this property is used to accommodate the range constraints of the problem. Indeed, the algorithms use the truncated normal distribution [22], which is confined to a given real interval $[a,b]$. The density function with mean μ and standard deviation $\sigma \geq 0$ is given by

$$f(x;\mu,\sigma,a,b) = \frac{\frac{1}{\sigma}\phi(\frac{x-\mu}{\sigma})}{\Phi(\frac{b-\mu}{\sigma}) - \Phi(\frac{a-\mu}{\sigma})},$$

and the cumulative distribution by

$$F(x;\mu,\sigma,a,b) = \frac{\Phi(\frac{x-\mu}{\sigma}) - \Phi(\frac{a-\mu}{\sigma})}{\Phi(\frac{b-\mu}{\sigma}) - \Phi(\frac{a-\mu}{\sigma})},$$

where ϕ and Φ denote the probability density and the cumulative distribution of the standard univariate normal, respectively.

13.3 Vine Approach in EDAs

This section provides a brief description of the C-vine and D-vine models and the motivation for using them to construct the search distributions of the EDAs. Furthermore, we introduce CVEDA and DVEDA, our third and fourth algorithms.

13.3.1 From Multivariate Copulas to Vines

Copulas model more realistic search distributions thanks to their ability to split joint densities into marginal information and a dependence structure. This overcomes the constraints imposed by the multivariate normal. However, the multivariate copula approach has also several shortcomings which have been clearly identified in the literature [1, 24]. Firstly, the number of tractable copulas when more than two variables are involved is rather limited. In fact the majority of the available parametric copulas are bivariate. Secondly, the multivariate copula approach is not appropriate when not all the pairs of variables have the same type of dependence structure. Finally, most of the existing multivariate copulas have only one parameter to quantify the overall dependence. This is a serious issue when the magnitude of dependence between pairs of variables is different.

An alternative to the multivariate copula approach is the pair-copula decomposition or vine [5, 6, 11]. The ability of vines to model a rich variety of dependencies by combining bivariate copulas of different families has motivated a growing research activity. This technique constructs high dimensional distributions using bivariate copulas as building blocks. When modeling with vines, an estimation procedure selects from a set of bivariate copulas, the one that fits best the data.

13.3.2 Vine Modeling Approach

Vines are dependency models of multivariate distribution functions based on the decomposition of $f(x_1,\ldots,x_n)$ into pair-copulas and marginal densities [1, 24]. A vine on n variables is a nested set of trees, $(T_1,\ldots,T_{n-1})$, where the edges of tree j are the nodes of the tree $j+1$ (with $j = 1,\ldots,n-1$) and each tree has the maximum number of edges. Regular vines constitute a special case of vine in which two edges in tree j are joined by an edge in tree $j+1$ only if these edges share a common node.

Two special types of regular vines are C-vines (canonical vines) and D-vines (drawable vines). Figure 13.1 shows a four-dimensional C-vine and D-vine. Each model has a specific way of decomposing the density. In particular, the C-vine density is given by

$$\prod_{k=1}^{n} f(x_k) \prod_{j=1}^{n-1} \prod_{i=1}^{n-j} c_{j,j+i|i,\ldots,j-1}\left(F\left(x_j|x_1,\ldots,x_{j-1}\right), F\left(x_{j+i}|x_1,\ldots,x_{j-1}\right)\right), \quad (13.3)$$

and the D-vine density is given by

$$\prod_{k=1}^{n} f(x_k) \prod_{j=1}^{n-1} \prod_{i=1}^{n-j} c_{i,i+j|i+1,\ldots,i+j-1}\left(F\left(x_i|x_{i+1},\ldots,x_{i+j-1}\right), F\left(x_{i+j}|x_{i+1},\ldots,x_{i+j-1}\right)\right), \quad (13.4)$$

where j identifies the trees and i denotes the edges in each tree.

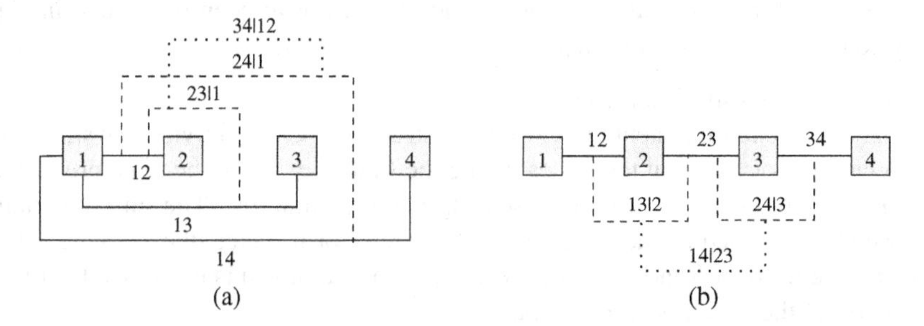

Fig. 13.1 Four-dimensional C-vine (a) and D-vine (b). In a C-vine, each tree T_j has a unique node that is connected to $n-j$ edges. In a D-vine, no node in any tree is connected to more than two edges

Note that in (13.3) and (13.4) the pair-copulas are evaluated at conditional distributions of the form $F(x \mid \mathbf{v})$. In [20] is showed that

$$F(x \mid \mathbf{v}) = \frac{\partial C_{xv_j|\mathbf{v}_{-j}}\left(F(x \mid \mathbf{v}_{-j}), F(v_j \mid \mathbf{v}_{-j})\right)}{\partial F(v_j \mid \mathbf{v}_{-j})},$$

where $C_{xv_j|\mathbf{v}_{-j}}$ is a bivariate copula distribution function, $\mathbf{v}$ is a n-dimensional vector, v_j is one component of $\mathbf{v}$ and $\mathbf{v}_{-j}$ denotes the $\mathbf{v}$-vector excluding the j component. The recursive evaluation of $F(x \mid \mathbf{v})$ yields the expression

$$F(x \mid v) = \frac{\partial C_{xv}\left(F_x(x), F_v(v)\right)}{\partial F_v(v)}.$$

When x and v are uniform, $F(x \mid v)$ reduces further to $F(x \mid v) = \partial C_{xv}(x,v)/\partial v$. Since the bivariate copulas may belong to different distribution families, the $h-$function,

$$h(x,v,\theta) = F(x \mid v) = \frac{\partial C_{xv}(x,v,\theta)}{\partial v},$$

is defined to facilitate de computation of $F(x \mid v)$, where θ denotes the set of parameters for the copula of the joint distribution function of x and v. Moreover, h^{-1} is the inverse of h with respect to the first variable (the inverse of $F(x \mid v)$). The derivation of h^{-1} for different bivariate copulas can be found in [1].

13.3.3 *Vine Estimation of Distribution Algorithm*

Vine Estimation of Distribution Algorithm (VEDA) [15, 36] is a class of EDA that uses vine-copula to model the search distributions. CVEDA and DVEDA are VEDAs based on C-vine and D-vine, respectively. Next we describe the particularities of the estimation and simulation steps of these algorithms.

13.3.3.1 Estimation

The methods for the construction of C- and D-vines have been developed in [1]. They consist of the following steps:

1. Selection of a specific factorization.
   In our current implementation, when constructing a C- and D-vine, the strongest relationships between the nodes define the first tree. Then, the structure of the subsequent trees can be defined starting from the first tree. The sum of certain weights assigned to each edge of the first tree (for instance, the absolute value of the empirical Kendall's τ between a pair of variables) is used by a heuristic to select the decomposition. In particular:

   - The construction of a C-vine begins by determining the weights between each variable (the possible root) and the others. The node of the tree that maximizes the sum of the weights of its edges is chosen as the root node of the first tree. In the subsequent trees, any edge in T_j is chosen as the root in T_{j+1}.
   - In a D-vine the final structure is uniquely determined by the structure of the first tree. The problem of constructing the first tree consists in finding the maximum weighted sequence (chain) of the variables. In [9] this problem is transformed into a traveling salesman problem (TSP) instance by adding a dummy node with weight zero on all edges to the other nodes. For efficiency reasons, we find an approximate solution of the TSP using the cheapest insertion heuristic [18, 31].

   The cost of the construction of these models increases with the number of variables. A n-dimensional C- or D-vine has $n-1$ trees and requires fitting copulas in $n(n-1)/2$ edges. However, the number of trees can be reduced if conditional independence is assumed or detected. If X and Y are conditionally independent given $\mathbf{V}$, then $c_{xy|\mathbf{v}}\left(F_{x|\mathbf{v}}(x \mid \mathbf{v}), F_{y|\mathbf{v}}(y \mid \mathbf{v})\right) = 1$. To simplify the construction of the model we apply the truncation strategy presented in [9, 10]. If a vine is truncated at a given tree, all the pair-copulas in the subsequent trees are assumed to be independence copulas. To identify the most appropriate truncation level, a greedy model selection procedure based on Akaike Information Criterion (AIC) [2] is used: The tree T_{j+1} is expanded if the AIC value calculated up to tree T_{j+1} is smaller than the AIC value obtained up to the previous tree; otherwise, the vine is truncated at T_j.

2. Choice of the pair-copula types in the factorization and estimation of the copula parameters.

    a. Determine the pair-copulas in the first tree from the original data.
    b. Compute observations (the conditional distribution functions of the form $F(x|y)$) for the second tree according to the h functions of the copulas in the first tree.
    c. Determine the pair-copulas in the second tree from observations obtained in step b.
    d. Repeat steps b and c for the following trees.

    At this step, a goodness-of-fit test (see [36]) is used to look for an appropriate fit among several candidate bivariate copulas: independence, normal, t, Clayton, rotated Clayton, Gumbel and rotated Gumbel. In this work, only the product and normal copulas are chosen. We apply a powerful nonparametric test of independence based on a CramÃľr-von Mises statistic [13]. The product copula is selected if there is not enough evidence against the null hypothesis of independence at a significance level of 0.01; otherwise, the normal copula is fitted.

13.3.3.2 Simulation

The simulation procedure of the C- and D-vines is based on the general sampling algorithm for n independent Uniform$(0,1)$ variables, allowing us to create a sample from the joint distribution function. The algorithm first samples n independent uniform random numbers $w_i \in (0,1)$, and then computes $x_1, x_2, \ldots, x_n$ according to $x_1 = w_1$, $x_2 = F_{2|1}^{-1}(w_2|u_1), \ldots, x_n = F_{n|1,2,\ldots,n-1}^{-1}(w_n|u_1,\ldots,u_{n-1})$.

13.4 Molecular Docking with VEDA

Molecular docking is a computational procedure to predict the geometry of binding of two molecules. Often, one of these molecules, the "receptor", is a protein, while the second one, the "ligand", is a small molecule that binds into the protein active site, usually a pocket or a groove on the protein surface. The protein-ligand docking problem remains open, since the algorithms for exploring the conformational space and the scoring functions that have been implemented so far, still have significant limitations [41]. In this work we address the conformational sampling problem.

In the chapter the protein is treated as a rigid body and the ligand as fully flexible. Therefore, the individuals represent the position, orientation and torsion angles of the ligand. The first three variables account for the ligand position in the three dimensional space. The 3D sampling is limited by a box enclosing the receptor binding site. For each test system, this box was constructed based on the minimum and maximum values of the ligand coordinates in its crystal conformation, plus a padding distance of $\pm 5A$ in each space dimension.

To represent the ligand orientation, we use three variables, the Euler angles, α, β, γ, that take values in $[0, 2\pi]$, $[-\pi/2, \pi/2]$ and $[-\pi, \pi]$, respectively. Thus, the position and orientation are defined using in total six variables. Finally, each flexible torsion angle in the ligand accounts for one variable, which takes values in $[-\pi, \pi]$. Hence the dimension of the optimization problem is six plus the number of torsion angles.

We use the semiempirical scoring function implemented in Autodock 4.2 [19]. The overall docking energy of a given ligand molecule is expressed as the sum of the pairwise interactions between receptor and ligand atoms (the intermolecular interaction energy) and the pairwise interactions between ligand atoms (the ligand intramolecular energy) as given by the expression

$$E = \Delta G_{vdv} E_{vdw} + \Delta G_{hbond} E_{hbond} + \Delta G_{elec} E_{elec} + \Delta G_{solv} E_{solv}.$$

The first three terms consider dispersion/repulsion, hydrogen bonding and the electrostatic potential. The last term accounts for desolvation effects. All terms are scaled empirically. To save calculation time the scoring function is evaluated using grids defined within the box enclosing the binding site, which account for the contributions from receptor atoms.

To evaluate the quality of the predicted ligand conformations, they are compared with the experimental crystal structure using the Cartesian root-mean-square deviation, $RMSD = \sqrt{\Sigma_{i=1}^{n} (dx_i^2 + dy_i^2 + dz_i^2)/n}$, where n is the number of ligand atoms and dx_i, dy_i and dz_i are the deviations between the crystallographic and predicted ligand coordinates of the ith atom. A structure with an $RMSD$ within $2A$ is classified as successfully docked, while a structure with an $RMSD$ between 2 and $3A$ is classified as partially docked.

13.5 Experiments

This section presents the experimental design and reports the numerical results of our empirical investigation.

13.5.1 Experimental Design

Table 13.1 presents the four protein-ligand complexes used as test systems, solved by X-ray crystallography and available from the Protein Data Bank (PDB) [8]. They were selected taking into account the large number of torsion angles in the ligand molecule.

We first find for each algorithm the population size that yields the lowest energy with the smallest number of evaluations, which ensures a comparison between the algorithms as fair as possible. To accomplish this, each algorithm was run 30

Table 13.1 Description of the test systems used in the experiments.

PDB code	Protein receptor	Number of ligand atoms*	Box dimensions (A)	Number of ligand torsions
1adb	Alcohol dehydrogenase	56	$28 \times 20 \times 22$	15
1bmm	Alpha-Thrombin	43	$17 \times 19 \times 22$	10
1cjw	Serotonin N-acetyltransferase	71	$26 \times 22 \times 30$	26
2z5u	Lysine-specific histone demethylase 1	73	$28 \times 32 \times 24$	20

* Non-polar hydrogens are not counted.

times using population sizes in the range between 200 and 2000 individuals with an increment of 200.

The EDAs stop when the standard deviation of the energy in the population is less than 0.01. It is worth noting that the optimal energy values for the test problems are unknown. The algorithms use a truncation selection of 0.3 and no elitism.

At this point, it is worth noting that we are not interested in the study of any sort of memetic strategy. We are aware of the possibilities of such methods and recommend their use in practical applications. However, for the sake of clarity of our results, we concentrate ourselves in the global exploration capabilities of the algorithms.

The algorithms were implemented using two packages programmed in R: `copulaedas` [16] and `vines` [17], which provide tools to build EDAs based on copula functions.

13.5.2 Numerical Results and Discussion

In this section we analyze the relative performance of the algorithms. Also, we study how the number of bivariate normal copulas fitted varies during the evolution of vine-based EDAs. Finally, several considerations about the adequacy of the fitness function model are given.

Table 13.2 summarizes the results obtained for each test system and each algorithm in terms of lowest energy ("best" solution), the number of function evaluations and the *RMSD* of the best solution with respect to the experimental conformation. The numbers shown in the table are the average values and their standard deviations over 30 independent runs. Figure 13.2 illustrates how the minimum energy decreases as the number of function evaluations increases during the optimization process. As shown in Table 13.2 and Figure 13.2, GCEDA, CVEDA and DVEDA perform better than UMDA, except for 1bmm. On the other hand, DVEDA and GCEDA behave slightly better than CVEDA. In general, the vine-based EDAs are able to perform satisfactorily both in problems with weak and strong interactions between the variables: While UMDA assumes independence and GCEDA postulates a normal dependency structure, CVEDA and DVEDA choose the bivariate copula that best fits the data; in this particular application, either an independence or a normal copula is chosen. The example of Figure 13.3 shows the correlations in the selected populations at generations 60 and 90 of DVEDA runs with 1bmm and 2z5u. These

Table 13.2 Average performance of copula-based EDAs.

PDB code	Algorithm	Population	Evaluations	Lowest Energy	*RMSD*
1adb	UMDA	1800	157933 ± 11286	−16.55 ± 2.09	3.17 ± 0.76
	CVEDA	1400	114258 ± 17585	−17.01 ± 2.18	2.63 ± 0.98
	DVEDA	1400	112133 ± 13920	−17.61 ± 1.79	2.50 ± 1.03
	GCEDA	1400	108200 ± 20162	−18.69 ± 0.84	1.44 ± 0.81
1bmm	UMDA	600	48633 ± 6099	−9.42 ± 1.37	4.58 ± 0.40
	CVEDA	800	65766 ± 10500	−9.11 ± 1.39	4.88 ± 0.49
	DVEDA	800	62266 ± 9776	−9.41 ± 1.33	4.77 ± 0.71
	GCEDA	1000	72166 ± 18632	−8.31 ± 1.10	4.99 ± 0.90
1cjw	UMDA	1200	180000 ± 13341	−12.19 ± 1.46	6.24 ± 1.04
	CVEDA	1200	180233 ± 13103	−12.85 ± 1.69	6.35 ± 1.33
	DVEDA	1200	165400 ± 18872	−14.40 ± 2.40	5.20 ± 1.34
	GCEDA	1600	201033 ± 37696	−15.55 ± 2.02	4.97 ± 1.26
2z5u	UMDA	1400	171900 ± 11442	−29.14 ± 1.97	0.61 ± 0.18
	CVEDA	1400	157600 ± 11391	−29.58 ± 1.23	0.58 ± 0.12
	DVEDA	1200	125266 ± 11965	−30.16 ± 1.28	0.52 ± 0.12
	GCEDA	1600	140966 ± 17835	−29.43 ± 0.56	0.51 ± 0.05

are the average generations where DVEDA fitted more normal copulas in these test systems (see Figure 13.4). In 1bmm, only weak correlations are observed, while in 2z5u, strong correlations between several pairs of variables are evident.

We now turn our attention to a comparison of the vine-based EDAs in terms of the relative proportion between the number of bivariate normal copulas that have been fitted and the total number of edges in the vine. Figure 13.4 presents the results for the 1bmm and 2z5u test systems. The number of bivariate copulas fitted by the algorithms increases during the evolution for the 2z5u system. For 1bmm the same behaviour up to the generation 60 is observed, and then decreases. It is interesting to note that after this point CVEDA fits slightly more normal copulas than DVEDA. Otherwise DVEDA fits more normal copulas than CVEDA, which is particularly significant in 2z5u. Indeed, although the construction procedure of the C-vine intends to represent explicitly the strongest correlations in the first tree, the constraint that only one variable can be connected to all the others may prevent some strong correlations to be included. As has been emphasized in [1], D-vines allow a more flexible selection of the dependencies to be explicitly modeled, while C-vines might be more appropriate in situations where one of the variables governs the interactions.

It is worth noting that because of the use of a truncation procedure (see Section 13.3.3.1) the number of statistical tests was dramatically reduced. In 2z5u the average number of fitted trees was below eight with CVEDA and below ten

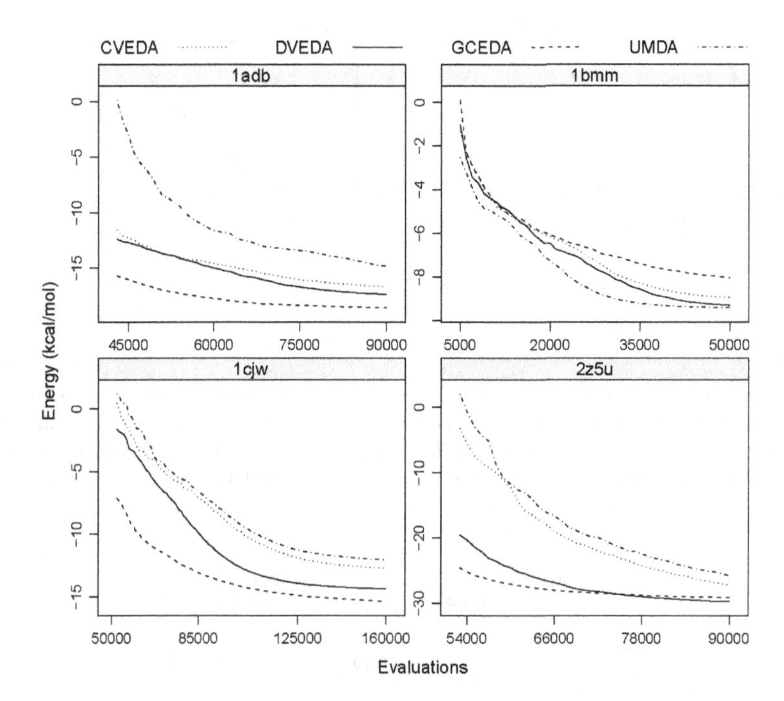

Fig. 13.2 Energy *versus* number of function evaluations during the evolution in the four protein-ligand test systems

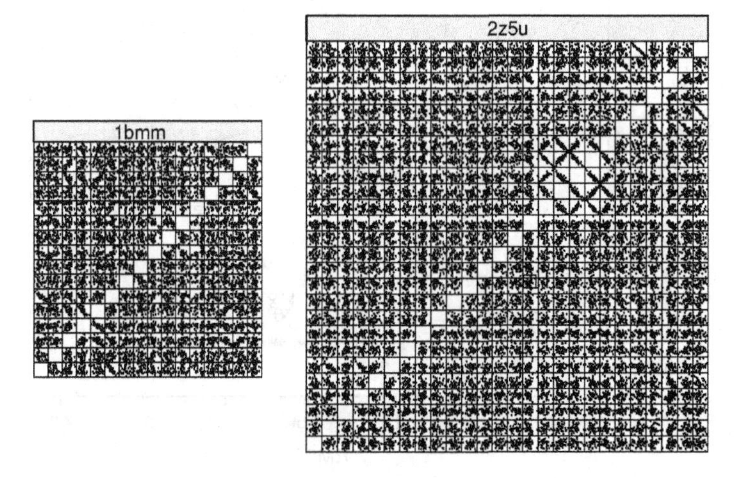

Fig. 13.3 Correlations in selected populations with DVEDA in 1bmm and 2z5u at generations 60 and 90, respectively. In 1bmm, weak correlations are observed, while in 2z5u, strong correlations between several pairs of variables (torsion angles) are evident

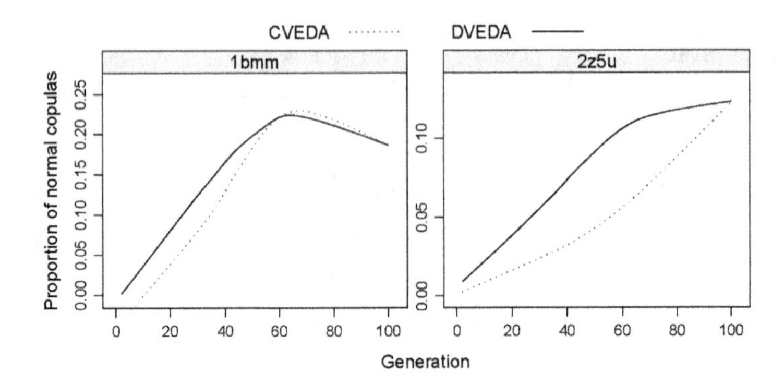

Fig. 13.4 Comparison of CVEDA and DVEDA for the 1bmm and 2z5u test systems, in terms of the relative proportion between the number of fitted normal copulas and the number of arcs in the vine, at different generations. For 1bmm, there are 15 trees and 120 arcs in a 16-dimensional C- and D-vine. For the 2z5u, there are 25 trees and 325 arcs in a 26-dimensional C- and D-vine

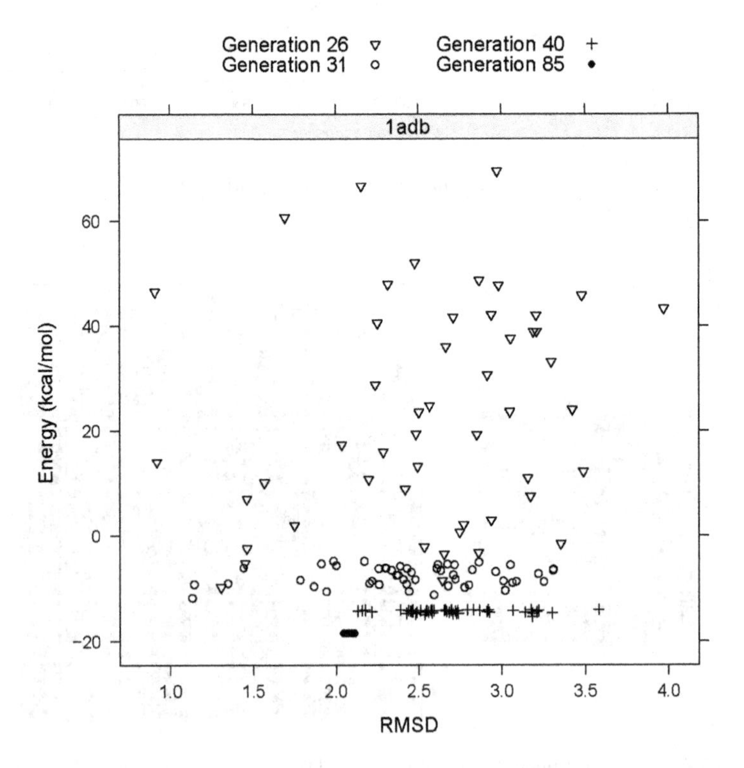

Fig. 13.5 Energy *versus RMSD* for the best solutions generated by DVEDA at different generations for the 1adb system. Solutions with higher (worse) energies may have better *RMSD*

with DVEDA, while in 1bmm, only a maximum of nine trees were fitted by both algorithms.

The ultimate test for a docking algorithm concerns the ability to reproduce the experimentally observed ligand conformations. This is usually evaluated in terms of the *RMSD* between the predicted and the crystal ligand coordinates, and the results heavily depend on the scoring function. Although such type of analysis involving the scoring function is outside the scope of this investigation, we illustrate here, in Figures 13.5 and 13.6, the results obtained from this test for the 1adb system.

Figure 13.5 shows a plot of energy versus *RMSD* for the best solutions generated by DVEDA at generations 26, 31, 40 and 85. Because of the limitations of the scoring function, there is no concordance between the energy scores and the ligand *RMSD*, so that solutions with higher energies may have better *RMSD*. Figure 13.6 illustrates such a situation for the ligand molecule in 1adb. The right half of the ligand is buried in the protein binding site, so its conformation is very restricted by the surrounding protein atoms. On the contrary, the left half of the ligand is exposed to the solvent and therefore is less affected by the steric constraints imposed by the protein (and so by the scoring function). The best solution at generation 85 (b), although it has a higher overall *RMSD*, shows a better fit for the buried part (right half), which explains its better energy as compared to solution at generation 26 (a).

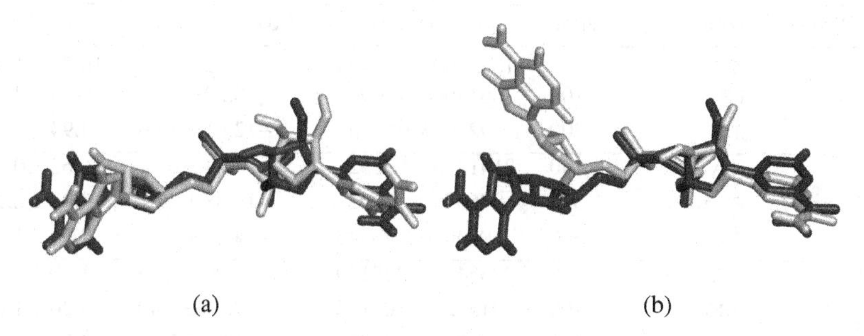

Fig. 13.6 Comparison of the known X-ray crystal structure of the 1adb (in black) with two predicted structures generated by DVEDA (in white): (a) At generation 26 the solution with the smallest *RMSD* of 0.8*A* has an energy value of 48.7 kcal/mol. (b) At generation 85 the solution with the smallest energy of 17.6 kcal/mol has an *RMSD* of 2.5*A*

13.6 GA, PSO and DE in the Molecular Docking Problem

In this section we include, for comparison with the EDAs, the results obtained with simple versions of a genetic algorithm (GA) [14] and a particle swarm optimization algorithm (PSO) [23]. These two population-based algorithms have been used in molecular docking applications [26, 29]. We also consider a differential evolution algorithm (DE) [38].

Similarly, as we did for the EDAs, we determined the population size that yields the lowest energy with the smallest number of evaluations. Each algorithm was run 30 times using population sizes in the range between 20 and 200 individuals with an increment of 20 for GA, and between 50 and 400 individuals with an increment of 50 for PSO and DE.

For GA [27] we use a two-point crossover rate of 0.8, a mutation based on Cauchy distribution with parameters $\alpha = 0$ and $\beta = 1$, a mutation rate of 0.2, an elitism value of one, and proportional selection. For PSO (Standard PSO 2007 [4]) we use an inertia weight of $1/2log(2)$, the user defined parameters are $\phi_1 = \phi_2 = 5log(2)$, and the number of neighborhoods is $1-(1-1/s)$ where s is the number of particles (swarm size). For DE (variant: DE/local-to-best/1/bin) we use a crossover rate of 0.8, a differential mutation rate of 0.5, and a mutation scale factor of 0.8.

The GA stops after a maximum number of 2.5 million energy evaluations, while PSO and DE stop after one million evaluations (the alternative stopping condition: if the standard deviation of the energy in the population is less than 0.01, is never achieved by these algorithms in the test suite used).

A comparison of the results shown in Table 13.2 with those in Table 13.3, confirms that copula-based EDAs outperform GA, PSO and DE, which is particularly significant for the GA. Only in 1bmm, PSO and DE obtain better energies than EDAs; however, it is worth noting the big difference on the number of evaluations.

Table 13.3 Average performance of GA, PSO and DE.

PDB code	Algorithm	Population	Evaluations	Lowest Energy	*RMSD*
	GA	20	1808000±432005	−8.72±5.30	10.71±1.91
1adb	PSO	100	863366±140913	−8.20±1.54	9.19±2.23
	DE	200	921073±93455	−12.61±1.66	4.94±2.47
	GA	140	2141200±301527	−8.84±0.81	6.62±0.92
1bmm	PSO	300	936630±132021	−11.88±2.06	3.98±2.25
	DE	300	933060±78343	−14.40±1.29	2.69±1.98
	GA	40	2235400±225130	−11.09±1.79	11.00±2.89
1cjw	PSO	100	915123±101167	−7.39±1.49	9.20±1.68
	DE	100	955600±48426	−12.94±0.52	7.28±1.57
	GA	20	2265600±273062	+483.50±538.39	13.34±4.21
2z5u	PSO	300	912510±95512	42.52±27.39	6.40±2.95
	DE	100	972536±47549	9.79±21.48	7.49±4.19

13.7 Conclusions

We have described an EDA approach to protein docking, which is based on undirected graphical models that use copulas. The empirical studies carried out on a test set of experimentally determined structures of protein-ligand complexes suggest that this approach performs competitively with the solutions of the state-of-the-art.

We have found that vine-based EDAs constitute promising tools for designing molecular docking algorithms, since they are endowed with mechanisms for building search distributions with different dependencies and with normal and non-normal margins. Vine-based EDAs are indeed more powerful and flexible than their predecessors UMDA and GCEDA. The results also suggest that the use of vines in EDAs open new opportunities to more appropriate modeling of the search distributions.

References

1. Aas, K., Czado, C., Frigessi, A., Bakken, H.: Pair-copula constructions of multiple dependence. Insurance: Mathematics and Economics 44, 182–198 (2009)
2. Akaike, H.: A new look at statistical model identification. IEEE Transactions on Automatic Control 19, 716–723 (1974)
3. Armañazas, R., Inza, I., Santana, R., Saeys, Y., Flores, J.L., Lozano, J.A., van de Peer, Y., Blanco, R., Robles, V., Bielza, C., Larrañaga, P.: A review of estimation of distribution algorithms in bioinformatics. BioData Mining 1(6) (2008)
4. Auger, A., Blackwell, T., Bratton, D., Clerc, M., Croussette, S., Dattasharma, A., Eberhart, R., Hansen, N., Keko, H., Kennedy, J., Krohling, R., Langdon, W., Li, W., Liu, A., Miranda, V., Poli, R., Serra, P., Stickel, M.: Standard PSO (2007), `http://www.particleswarm.info/`
5. Bedford, T., Cooke, R.M.: Probability density decomposition for conditionally dependent random variables modeled by vines. Annals of Mathematics and Artificial Intelligence 32, 245–268 (2001)
6. Bedford, T., Cooke, R.M.: Vines – a new graphical model for dependent random variables. The Annals of Statistics 30, 1031–1068 (2002)
7. Belda, I., Madurga, S., Llorá, X., Martinell, M., Tarragó, T., Piqueras, M., Nicolás, E., Giralt, E.: ENPDA: An evolutionary structure-based de novo peptide design algorithm. Journal of Computer-Aided Molecular Design 19(8), 585–601 (2005)
8. Berman, H.M., Westbrook, J., Feng, Z., Gilliland, G., Bhat, T.N., Weissig, H., Shindyalov, I.N., Bourne, P.E.: The protein data bank. Nucleic Acids Research 28, 235–242 (2000)
9. Brechmann, E.C.: Truncated and simplified regular vines and their applications. Diploma thesis, Technische Universität München (2010)
10. Brechmann, E.C., Czado, C., Aas, K.: Truncated regular vines in high dimensions with application to financial data. Note SAMBA/60/10, Norwegian Computing Center, NR (2010)
11. Cooke, R.M.: Markov and entropy properties of tree- and vine-dependent variables. In: Proceedings of the American Statistical Association Section on Bayesian Statistical Science, pp. 166–175 (1997)
12. Cuesta-Infante, A., Santana, R., Hidalgo, J.I., Bielza, C., Larrañaga, P.: Bivariate empirical and n-variate Archimedean copulas in estimation of distribution algorithms. In: Proceedings of the IEEE Congress on Evolutionary Computation (CEC 2010), pp. 1355–1362 (2010)
13. Genest, C., Rémillard, B.: Tests of independence or randomness based on the empirical copula process. Test 13, 335–369 (2004)
14. Goldberg, D.E.: Genetic Algorithms in Search, Optimization and Machine Learning. Addison-Wesley, Reading (1989)
15. González-Fernández, Y.: Algoritmos con estimación de distribuciones basados en cópulas y vines. Diploma thesis, University of Havana (June 2011)

16. González-Fernández, Y., Soto, M.: copulaedas: Estimation of Distribution Algorithms Based on Copula Theory, R package version 1.0.1. (2011), `http://CRAN.R-project.org/package=copulaedas`
17. González-Fernández, Y., Soto, M.: vines: Multivariate Dependence Modeling with Vines, package version 1.0.1., p. 1 (2011), `http://CRAN.R-project.org/package=vines`
18. Hahsler, M., Hornik, K.: TSP – Infrastructure for the traveling salesperson problem. Journal of Statistical Software 23, 1–21 (2007)
19. Huey, R., Morris, G.M., Olson, A.J., Goodsell, D.S.: A semiempirical free energy force field with charge-based desolvation. Journal Computational Chemistry 28, 1145–1152 (2007)
20. Joe, H.: Families of m-variate distributions with given margins and $m(m-1)/2$ bivariate dependence parameters. In: Distributions with Fixed Marginals and Related Topics, pp. 120–141 (1996)
21. Joe, H.: Multivariate Models and Dependence Concepts. Chapman & Hall (1997)
22. Johnson, N.L., Kotz, S., Balakrishnan, N.: Continuous Univariate Distributions, 2nd edn., vol. 1. John Wiley & Sons (1994)
23. Kennedy, J., Eberhart, R.C.: Particle swarm optimization. In: Proceedings of the IEEE International Conference on Neural Networks IV, pp. 1942–1948 (1995)
24. Kurowicka, D., Cooke, R.M.: Uncertainty Analysis with High Dimensional Dependence Modelling. John Wiley & Sons (2006)
25. Larrañaga, P., Lozano, J.A. (eds.): Estimation of Distribution Algorithms. An New Tool for Evolutionary Computation. Kluwer Academic Publisher (2002)
26. Morris, G.M., Goodsell, D.S., Halliday, R.S., Huey, R., Hart, W.E., Belew, R.K., Olson, A.J.: Automated docking using a Lamarckian genetic algorithm and an empirical binding free energy function. Journal of Computational Chemistry 19(14), 1639–1662 (1998)
27. Morris, G.M., Goodsell, D.S., Pique, M.E., Lindstrom, W., Halliday, R.S., Huey, R., Forli, S., Hart, W.E., Belew, R.K., Olson, A.J.: Automated Docking of Flexible Ligands to Flexible Receptors. User Guide AutoDock. Version 4.2 (2010)
28. Mühlenbein, H., Paaß, G.: From Recombination of Genes to the Estimation of Distributions I. Binary Parameters. In: Ebeling, W., Rechenberg, I., Voigt, H.-M., Schwefel, H.-P. (eds.) PPSN 1996. LNCS, vol. 1141, pp. 178–187. Springer, Heidelberg (1996)
29. Namasivayam, V., Günther, R.: Flexible peptide-protein docking employing pso@autodock. In: From Computational Biophysics to Systems Biology (CBSB 2008), vol. 40, pp. 337–340 (2008)
30. Nelsen, R.B.: An Introduction to Copulas, 2nd edn. Springer (2006)
31. Rosenkrantz, D.J., Stearns, R.E., Lewis, P.M.: An analysis of several heuristics for the traveling salesman problem. SIAM Journal on Computing 6(3), 563–581 (1977)
32. Rousseeuw, P., Molenberghs, G.: Transformation of nonpositive semidefinite correlation matrices. Communications in Statistics: Theory and Methods 22, 965–984 (1993)
33. Santana, R.: Advances in Probabilistic Graphical Models for Optimization and Learning. Applications in Protein Modeling. PhD thesis, University of the Basque Country (2006)
34. Santana, R., Larrañaga, P., Lozano, J.A.: Side chain placement using estimation of distribution algorithms. Artificial Intelligence in Medicine 39, 49–63 (2007)
35. Sklar, A.: Fonctions de répartition à n dimensions et leurs marges. Publications de l'Institut de Statistique de l'Université de Paris 8, 229–231 (1959)
36. Soto, M., González-Fernández, Y.: Vine estimation of distribution algorithms. Technical Report ICIMAF 2010-561, Institute of Cybernetics, Mathematics and Physics (May 2010) ISSN 0138-8916

37. Soto, M., Ochoa, A., Arderí, R.J.: Estimation of distribution algorithm based on Gaussian copula. Technical Report ICIMAF 2007-406, Institute of Cybernetics, Mathematics and Physics (June 2007) ISSN 0138-8916
38. Storn, R., Price, K.: Differential evolution – a simple and efficient heuristic for global optimization over continuous spaces. Journal of Global Optimization 11, 341–359 (1997)
39. Wang, L.F., Wang, Y., Zeng, J.C., Hong, Y.: An estimation of distribution algorithm based on Clayton copula and empirical margins. In: Life System Modeling and Intelligent Computing, pp. 82–88. Springer (2010)
40. Wang, L.F., Zeng, J.C., Hong, Y.: Estimation of distribution algorithm based on copula theory. In: Proceedings of the IEEE Congress on Evolutionary Computation (CEC 2009), pp. 1057–1063 (2009)
41. Warren, G.L., Andrews, C.W., Capelli, A.M., Clarke, B., LaLonde, J., Lambert, M.H., Lindvall, M., Nevins, N., Semus, S.F., Senger, S., Tedesco, G., Wall, I.D., Woolven, J.M., Peishoff, C.E., Head, M.S.: A critical assessment of docking programs and scoring functions. Journal of Medicinal Chemistry 49, 5912–5931 (2006)

Chapter 14
EDA-RL: EDA with Conditional Random Fields for Solving Reinforcement Learning Problems

Hisashi Handa

Abstract. This chapter introduces a novel Estimation of Distribution Algorithm for solving Reinforcement Learning Problems, i.e., EDA-RL. As the probabilistic model of the EDA-RL, the Conditional Random Fields proposed by Lafferty *et al.* are employed. The Conditional Random Fields can estimate conditional probability distributions by using Markov Network. Moreover, the structural search of probabilistic model by using χ^2-test, and data correction method are examined. One of the primary features of the EDA-RL is the direct estimation of reinforcement learning agents' policies by using the Conditional Random Fields. Another feature is that a kind of undirected graphical probabilistic model is used in the proposed method. The experimental results on Probabilistic Transition Problems and Maze Problems show the effectiveness of the EDA-RL.

14.1 Introduction

In this chapter, a novel Estimation of Distribution Algorithm for solving reinforcement learning problems, called EDA-RL, is introduced. In the EDA-RL, instead of individuals, input-output records (episodes) generated by the interaction between agents and environments are stored. Better episodes are selected and are then used to estimate the policies of agents. The Conditional Random Fields (CRFs) — a sort of Markov Network used for estimating conditional probability distributions — are used to estimate the policy of agents. Furthermore, new episodes are generated by agents with these new policies, which are estimated from better episodes in the previous generation. Therefore, in the EDA-RL, the probabilistic model can be regarded as a solution, whereas most EDAs use individuals to represent solutions. That is, the objective of EDA-RL is to learn a probabilistic model which capture the correct agent-environment interactions.

Hisashi Handa
Okayama University, Okayama 700-8530, Japan
e-mail: handa@sdc.it.okayama-u.ac.jp

S. Shakya and R. Santana (Eds.): Markov Networks in Evolutionary Computation, ALO 14, pp. 227–239.
springerlink.com

Conventional discrete Reinforcement Learning Algorithms provide us with an algorithmic tool to determine agents with optimal policies using a Markov Decision Process [9][10]. In practice, however, such a theoretical base may be inapplicable since some problems are far from being Markov Decision Processes. In addition, a conventional Reinforcement Learning Algorithm is a kind of local search algorithm, where reward and value information is locally propagated into neighbor states. Therefore, in the case of huge problems, such value propagation results in considerable time being taken for the whole problem space to be learnt or alternatively, the learning cannot be done well.

In the case of games and autonomous robot navigation, evolving neural networks and evolutionary fuzzy systems have attracted much attention recently [7][4]. In such approaches, evolutionary computation optimizes parameters in controllers such as neural networks or fuzzy systems. In this case, genetic operators are applied to a parameter space so that modifications made by these operators are not taken account of in the effects on agents' policies.

Other EDA approach for reinforcement learning problems is the use of Bayesian Optimisation Algorithm (BOA) for rule extraction in in accuracy-based classifier system (XCS) [1, 2]. Individuals, i.e. rules, in this approach also represent solution as in most EDAs. The scheme using probabilistic model in this approach is to use better reproduction mechanism, as in conventional EDAs.

The organization of the remainder of this chapter is as follows: Section 14.2 begins by introducing Reinforcement Learning Problems. Section 14.3 briefly summarizes Conditional Random Fields. The EDA-RL is introduced in section 14.4 by referring to several probabilistic graphical models, the structure search of probabilistic model, and data correction method. Computer simulations of Probabilistic Transition Problems and maze problems are carried out in Section 14.5.

14.2 Reinforcement Learning Problems

The reinforcement learning problem is the problem of learning from interactions with an environment. Such interactions are composed of perceptions of the environment and actions that affect both agents and environments. Agents try to maximize the total reward received from the environment as consequences of their actions. In other words, the task for agents is to acquire a policy $\pi(s,a)$ which maximizes the prospective reward. The variables s and a in $\pi(s,a)$ denote states recognized by agents, and actions taken by agents, respectively. The policy can be defined by using a probabilistic formulation such as $P(a|s)$. Most Reinforcement Learning algorithms learn value functions, e.g., a state-action value function $Q(s,a)$ and a state value function $V(s)$, instead of estimating the policy $\pi(s,a)$ directly. That is, in the case of conventional Reinforcement Learning algorithms, the policy $\pi(s,a)$ is approximated by value functions. In the EDA-RL, we adopt Conditional Random Fields to estimate the policy $\pi(s,a)$ from selected episodes in the previous generation.

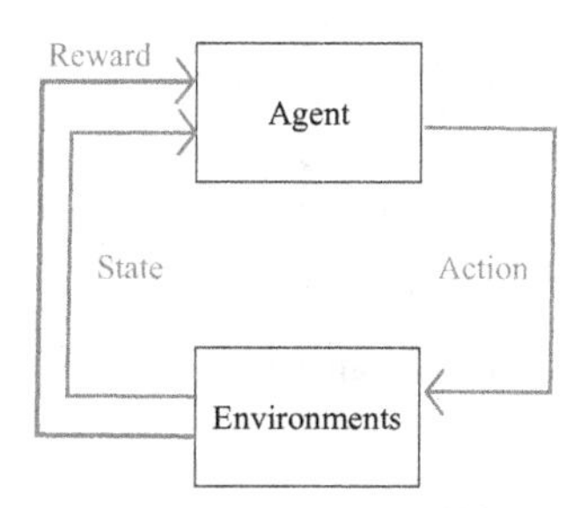

Fig. 14.1 Reinforcement Learning Problems

14.3 Conditional Random Fields

14.3.1 Overview

The Conditional Random Fields (CRFs) proposed by Lafferty *et al.* have been applied to segmentation problems in text processing and bio-informatics [5][8]. The CRFs can handle a conditional distribution $P(\mathbf{y}|\mathbf{x})$ with an associated graphical structure. A notable feature of the CRFs is that they model such conditional distributions whereas Hidden Markov Models, which are traditionally used in broad areas such as voice recognition and text processing, estimate joint probability distributions $P(\mathbf{x},\mathbf{y})$. This implies that we do not have to consider the probabilistic distribution of inputs $P(\mathbf{x})$ since $P(\mathbf{x},\mathbf{y}) = P(\mathbf{y}|\mathbf{x}) \cdot P(\mathbf{x})$. In general, the probability distribution of the inputs $P(\mathbf{x})$ is unknown. Moreover, in the case of Reinforcement Learning Problems, it depends on agents' actions.

CRFs can be regarded as a sort of Markov Network since CRFs use an undirected graph model to represent a probabilistic distribution. That is, variables in the problem are factorized in advance. Each clique (factor) is associated with a local function. In the case of a log-linear Markov Network, the joint probability $P(\mathbf{y},\mathbf{x})$ can be represented as follows:

$$P(\mathbf{y},\mathbf{x}) = \frac{1}{Z}\prod_{A}\Psi_A(\mathbf{y}_A,\mathbf{x}_A), \tag{14.1}$$

where Ψ_A denotes a local function for a set of variables $\mathbf{y}_A,\mathbf{x}_A \in \{\mathbf{y},\mathbf{x}\}$. Z is a normalizing factor which ensures that the distribution sums to 1: $Z = \sum_{\mathbf{y},\mathbf{x}}\prod_A \Psi_A(\mathbf{y}_A,\mathbf{x}_A)$.

Meanwhile, CRFs can handle the following conditional probabilities $P(\mathbf{y}|\mathbf{x})$:

$$\begin{aligned} P(\mathbf{y}|\mathbf{x}) &= \frac{P(\mathbf{y},\mathbf{x})}{\sum_{\mathbf{y}'} P(\mathbf{y}',\mathbf{x})} = \frac{\frac{1}{Z}\prod_A \Psi_A(\mathbf{y}_A,\mathbf{x}_A)}{\sum_{\mathbf{y}'}\frac{1}{Z}\prod_A \Psi_A(\mathbf{y}'_A,\mathbf{x}_A)} = \frac{\prod_A \Psi_A(\mathbf{y}_A,\mathbf{x}_A)}{\sum_{\mathbf{y}'}\prod_A \Psi_A(\mathbf{y}'_A,\mathbf{x}_A)} \\ &= \frac{1}{Z(\mathbf{x})}\prod_A \Psi_A(\mathbf{y}_A,\mathbf{x}_A), \end{aligned} \tag{14.2}$$

where $Z(\mathbf{x}) = \sum_{\mathbf{y}'}\prod_A \Psi_A(\mathbf{y}'_A,\mathbf{x}_A)$.

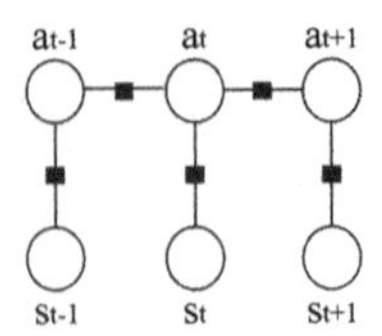

Fig. 14.2 Graphical model of a Linear-Chain CRF

14.3.2 Linear-Chain CRF

This subsection introduces the linear-chain CRF which is widely used in the natural text-processing area. The linear-chain CRF is one of the simplest CRFs. Here, input and output variables $\mathbf{x}$, $\mathbf{y}$ in the previous subsection are substituted by a sequence of states $\mathbf{s} = \{s_1, s_2, \ldots, s_T\}$ and a sequence of corresponding actions $\mathbf{a} = \{a_1, a_2, \ldots, a_T\}$, respectively. The linear-chain graphical model in this case is depicted in Figure 14.2. This depiction is called factor graphs. Each circle (node) in this figure represents a state or an action at the corresponding time step. Each line with a black square in the middle of the line indicate that nodes are associated with a local function Ψ.

As we can see from this figure, the linear-chain CRFs factorize an episode, i.e., a sequence of state-action pairs $(\mathbf{s}, \mathbf{a})$ into state-action pairs (s_t, a_t) and transitions of actions (a_{t-1}, a_t). The local function for each time step is defined as follows:

$$\Psi_k(\mathbf{a}, \mathbf{s}) = \begin{cases} \exp(\lambda_k \cdot u_k(a_t, s_t)) & k \leq K' \\ \exp(\lambda_k \cdot v_k(a_{t_{k-1}}, a_t)) & K' < k \leq K, \end{cases}$$

where $u_k(a_t, s_t)$ and $v_k(a_{t_{k-1}}, a_t)$ are feature functions and K' is the total number of possible combinations of states and actions (a, s). K is the total number of the feature functions. The values $u_k(a_t, s_t)$ and $v_k(a_{t_{k-1}}, a_t)$ are set to 1 if the corresponding state-action pair and action-action pair are observed, respectively. λ_k denotes a parameter for a factor k. Equation (14.2) can be rewritten by using the above notation as follows:

$$P(\mathbf{a}|\mathbf{s}) = \frac{1}{Z(\mathbf{s})} \exp\left\{ \sum_{t=1}^{T} \left\{ \sum_{k=1}^{K'} \lambda_k u_k(a_t, s_t) + \sum_{k=K'+1}^{K} \lambda_k v_k(a_{t-1}, a_t) \right\} \right\}. \quad (14.3)$$

14.3.3 Parameter Estimation

Suppose that N episodes $(\mathbf{s}^{(i)}, \mathbf{a}^{(i)})$, $(i = 1, \ldots N)$ are acquired to estimate the policy and each episode $(\mathbf{s}^{(i)}, \mathbf{a}^{(i)})$ is composed of a series of state-action pairs:

$$(\mathbf{s}^{(i)}, \mathbf{a}^{(i)}) = \{(s_1, a_1), (s_2, a_2), \ldots (s_{T_i}, a_{T_i})\},$$

where T_i denotes the length of episode i. The log likelihood method is used to estimate the parameters $\theta = (\lambda_1, \ldots, \lambda_K)$:

$$
\begin{aligned}
l(\theta) &= \sum_{i=1}^{N} \log P(\mathbf{a}^{(i)}|\mathbf{s}^{(i)}) \\
&= \sum_{i=1}^{N}\sum_{t=1}^{T_i}\left\{\sum_{k=1}^{K'} \lambda_k u_k(a_t, s_t) + \sum_{k=K'+1}^{K} \lambda_k v_k(a_{t-1}, a_t)\right\} - \sum_{i=1}^{N} \log Z(\mathbf{s}^{(i)})
\end{aligned}
$$

In order to calculate the optimal parameter θ, the partial derivative of the above equation is used. For $k \leq K'$, we can calculate the partial derivative as follows:

$$
\begin{aligned}
\frac{\partial l}{\partial \lambda_k} &= \sum_{i=1}^{N}\sum_{t=1}^{T_i} u_k(a_t, s_t) - \sum_{i=1}^{N} \frac{(Z(\mathbf{s}^{(i)}))'}{Z(\mathbf{s}^{(i)})} \\
&= \sum_{i=1}^{N}\sum_{t=1}^{T_i} u_k(a_t, s_t) - \sum_{i=1}^{N}\sum_{a'_t} P(a'_t|\mathbf{s}^{(i)}) \cdot u_k(a'_t, s_t).
\end{aligned}
$$

The first and second terms of the right hand side of this equation denote the number of observations of the factor $u_k(a_t, s_t)$ and the expected number of observations of the factor $u_k(a_t, s_t)$ under the current value of the parameter θ, respectively. Therefore, this derivative indicates that the parameter λ_k is modified along that expected value closest to the actual observation in the selected episodes. Further descriptions needed to calculate $l(\theta)$ and $\frac{\partial l}{\partial \lambda_k}$, i.e., $Z(\mathbf{s}^{(i)})$ and $P(a'_t|\mathbf{s}^{(i)})$ are described in [3].

14.4 EDA-RL

14.4.1 Calculation Procedure of EDA-RL

Figure 14.3 depicts the calculation procedure of the proposed method, i.e., EDA-RL. The procedure is summarized as follows:

1. The initial policy $\pi(s,a)(= P(a|s))$ is taken to be a uniform distribution. That is, according to the initial policy $\pi(s,a)$, agents move randomly.
2. Agents interact with the environment by using policy $\pi(s,a)$ until T_s episodes are generated.
3. The best episode, i.e., the episode with greatest reward, among the T_s episodes in the previous step is stored in the episode database. Return to step 2. until the number of chosen episodes reaches a predefined constant value C_d.
4. A new policy $\pi(s,a)$ for the set of episodes in the database is estimated by CRF. After the estimation, all episode data in the database is erased.
5. Return to step 2. until terminal conditions are met.

One of the main differences between conventional EDAs and EDA-RL is the use of the estimated probabilistic model. Conventional EDAs employ a probabilistic model to generate individuals, i.e., solutions for given problems. On the other hand, the probabilistic model estimated by CRF represents the policy $\pi(s,a)$. In other words, the probabilistic model denotes the solution itself.

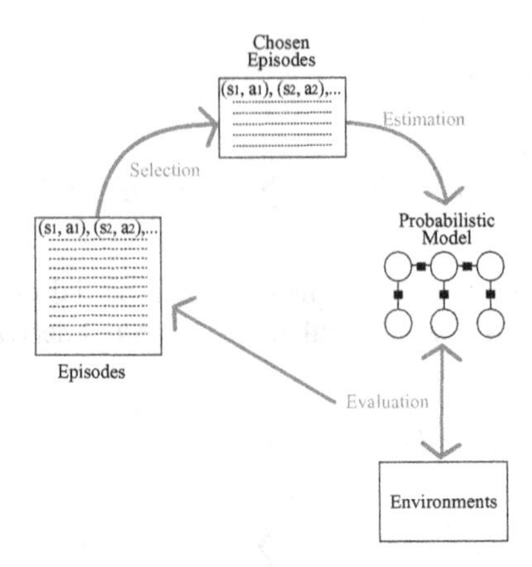

Fig. 14.3 Search Process of an EDA-RL

We note that in step 1., in the above procedure, we do not have to assume a uniform distribution for the initial policy. That is, if there is plenty of observation data available, e.g. play-data by humans, or episodes from the conventional approach, such data can be used to generate an initial policy. This means that the proposed method can easily incorporate human knowledge and can improve on it by using an evolutionary approach.

14.4.2 Probabilistic Models

In addition to the linear-chain CRF, two other probabilistic graphical models are examined in this chapter. These models are depicted in Figure 14.4 and can be described as follows:

- **Independent features:** This probabilistic model assumes that observations of an agent during an episode are independent. Therefore, factors $u_k(a_t, s_t)$ are used so that the following conditional probability is employed:

$$P(\mathbf{a}|\mathbf{s}) = \frac{1}{Z(\mathbf{s})} \exp\left\{ \sum_{t=1}^{T} \sum_{k=1}^{K} \lambda_k u_k(a_t, s_t) \right\}.$$

- **3-clique:** In this probabilistic model, two factors in the linear-chain probabilistic model are merged into one factor: $c_k(a_{t-1}, a_t, s_t)$. Therefore, this probabilistic model does not use $u_k(a_t, s_t)$ and $v_k(a_{t-1}, a_t)$:

$$P(\mathbf{a}|\mathbf{s}) = \frac{1}{Z(\mathbf{s})} \exp\left\{ \sum_{t=1}^{T} \sum_{k=1}^{K} \lambda_k c_k(a_{t-1}, a_t, st) \right\}.$$

As in the 3-Clique probabilistic model, larger cliques imply that the number of factors, i.e., possible combinations among states and actions, are increased if the

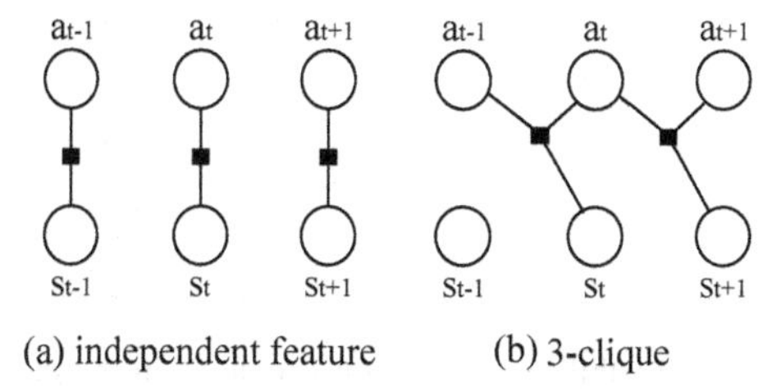

Fig. 14.4 Depiction of probabilistic models: (a) Independent Features, and (b) 3-Clique

number of states is greater than the number of actions. Suppose that there are n states in a problem to be solved, and agents can execute m actions in each state. In this case, the number of factors for the linear-chain probabilistic model is $nm + m^2$ whereas the number for the 3-clique is nm^2.

14.4.3 Structure Search of Probabilistic Models

Perceptual aliasing states in reinforcement learning problems stand for states where several different states are regarded as the same state by agents. In simpler cases, historical information can help for solving such difficulties. The utilization of such historical information means that agents use augmented state representation including preceding actions and/or states. In order to decide if certain states should use historical information, χ^2-test is employed. That is, before building probabilistic models at each generation, new augmented states are added as a consequence of the test as following equation:

$$\chi^2(s, a_{t-1}) = \frac{\Sigma_a (p(a|s, a_{t-1}) - p(a|s)p(a|a_{t-1}))}{(p(a|s)p(a|a_{t-1})},$$

where $p(a|s)$, $p(a|a_{t-1})$, and $p(a|s, a_{t-1})$ are the conditional probability of action a given by state s, the one of action a given by preceding action a_{t-1}, and the one of action a given by state s and preceding action a_{t-1}, respectively. These probabilities are calculated over the episode data in the database of step 3) in section 14.4.1. This calculation is carried out before building the probabilistic mode. For all the possible combinations of (s, a_{t-1}), the above function is calculated. New feature functions $c_l(a_{t-1}, a_t, s)$ are incorporated if $\chi^2(s, a_{t-1})$ exceeds a predefined threshold value. Hence, the independent feature model is used as an initial probabilistic model at first, and then these feature functions $c_l(a_{t-1}, a_t, s)$ are added to certain pairs of a state and a preceding action (s, a_{t-1}). As a consequence of this incorporation, the probabilistic model of agents is changed as follows:

$$P(\mathbf{a}|\mathbf{s}) = \frac{1}{Z(\mathbf{s})} \exp\left\{ \sum_{t=1}^{T} \left\{ \sum_{k=1}^{K} \lambda_k u_k(a_t, s_t) + \sum_{l=1}^{L} \lambda_l c_l(a_{t-1}, a_t, st) \right\} \right\}.$$

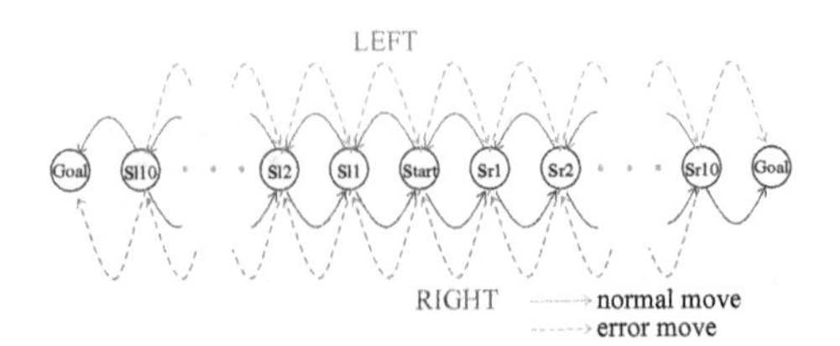

Fig. 14.5 A depiction of Probabilistic Transition Problems in the case where there are 10 intermediate states

14.4.4 Data Correction by Loop Elimination

As pointed out in [6], in the case of Profit Sharing[1], iterated state action pairs in episode data can be eliminated. In addition, such elimination helps agents to acquire rational policies. EDA-RL can be regarded as non-bootstraping as in the Profit Sharing method.

Loop elimination introduced in this chapter is carried out as follows:

1. For each time step t in episode e, states s_t^{pre} and s_t^{post} which are observed before and after the time step t, respectively.
2. Two (state, action)-pairs are found out in the episode e such that i) $s_{t_1} = s_{t_2}$, ii) $a_{t_1} = a_{t_2}$, iii) $s_{t_1}^{pre} = s_{t_2}^{pre}$, iv) $s_{t_1}^{post} = s_{t_2}^{post}$ and v) $t_2 - t_1$ is maximized, where t_1 and t_2 denote time steps pairs are observed.
3. This procedure is terminated if no pairs are found.
4. The episode data from the next of t_1 to t_2 are eliminated. Go back to step 2.

14.5 Experiments

14.5.1 Problem Settings

Probabilistic Transition Problems (PTP) and two Maze Problems are introduced in order to investigate the effectiveness of the proposed method.

An example of the PTP is depicted in Figure 14.5, a start point is located at the center position. Agents can take two actions at any states: a left or a right move. By taking these actions, agents move to an adjacent state. However, with probability P_w, agents move in the opposite direction to their chosen action; this is called an error move. There are two goals, one goal rewards an agent with $100/c$, where c denotes the number of steps until an agent reaches its goal. The other goal punishes an agent with a negative reward: $-100/c$. When agents reach one of the goals, the episode is finished. The number of intermediate states n_s can be changed. In the following experiments, we examined cases where the numbers of intermediate states were 20 and 40.

[1] The Profit Sharing updates the values of states or state-action pairs at the ends of episodes by distributing the total amount of reward based upon weights.

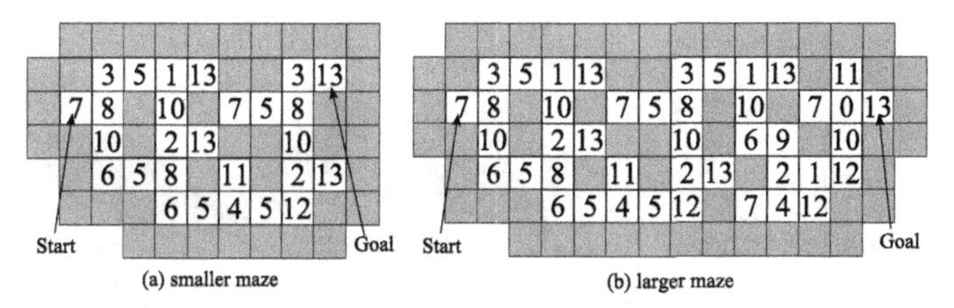

Fig. 14.6 Maze problems [11] examined in this chapter

Two maze problems as delineated in Figure 14.6 [11] are examined. Numbers in cells stand for states, which agents can recognize. These numbers correspond to the alignment of surrounding walls. The states of surrounding walls are represented by a sequence of binary string 1001. Then, such binary string is traslated into a digit number. For example, state “5” is assigned if there are two walls on the upper and lower sides. The max length of episodes is set to be 1000. Fitness, i.e., reward, for episodes is defined as 1000 - n_t, where n_t means the number of time steps for reaching goals. The number of time steps n_t is affected by agents’ policy, i.e., the estimated probabilistic model by using the CRF.

14.5.2 Experimental Results

We examined the three probabilistic models discussed in sections 14.3.2 and 14.4.2 with different database sizes C_d and different numbers of episodes T_s for each selection operation. C_d and T_s were set to be one of $\{16,32,64,128,256,512\}$ and one of $\{2,3,4,5,10\}$, respectively. The number of generations was set to be 30. However, this setting does not affect results since algorithms are allowed to converge at earlier generations if they can do so.

Figure 14.7 shows the averaged value of the acquired reward at the final generation for the PTP. The “tournament size” stands for T_s in the previous subsection. The average is calculated over all the episodes in the final generation and over 20 runs. This reward is plotted on the Z axis and is normalized to 1, i.e., such that the performance of the optimal policy is mapped to 1. In other words, the value 1 in each graph indicates that the algorithm with corresponding parameter settings can find the optimal policy for all runs. From the upper to lower rows, there are the results for $(n_s, P_w) = (20,0.1)$, $(20,0.3)$, $(40,0.3)$ for the probabilistic transition problems.

The results for the probabilistic transition problems in Figure 14.7 tell us that the linear chain model shows good performance. The structure of the probabilistic model in the linear chain model may be suitable for these problems. The independent features and 3-clique models were able to work well if there were sufficient good episode data. That is, the number of selected individuals C_d and larger number of episodes T_s for each selection operation have a considerable effect on these methods.

Figures 14.8 and 14.9 show the experimental results for the smaller and the larger maze problems, respectively. In these maze problems, the EDA-RL with structural

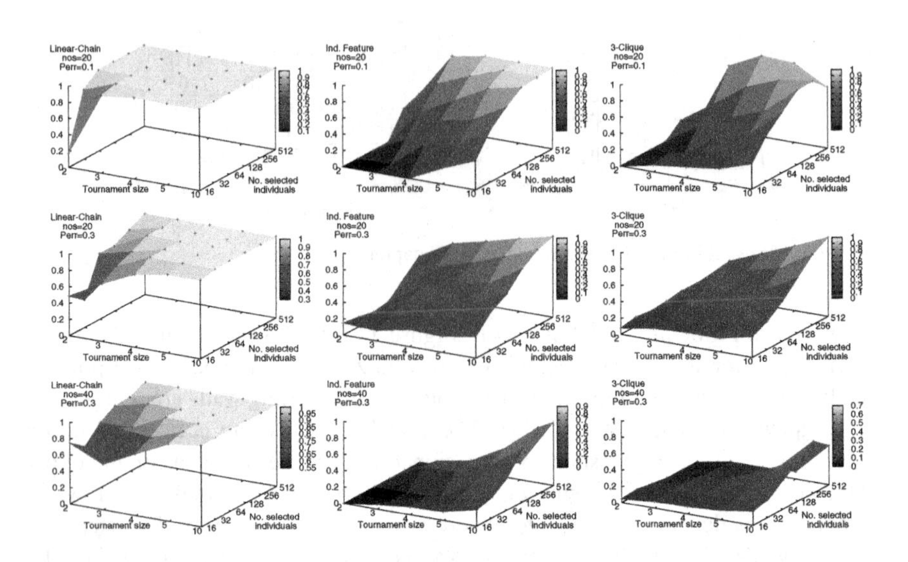

Fig. 14.7 Experimental results for the probabilistic transition problems: normalized acquired reward; These results are averaged over 20 runs

search is compared with the linear-chain model and 3-clique model. Experiments are done with different number of episodes for policy estimation C_d and different numbers of episodes T_s for each selection operation. C_d and T_s were set to be one of $\{64, 128, 256, 512\}$ and one of $\{2, 5, 10, 20\}$, respectively. Table 14.1 summarizes the number of features of the proposed method at the end of evolution. The left, center, and right graphs in these figures denote the result of EDA-RL with linear-chain model, with 3-clique model, and with structural search method, respectively. The upper and lower graphs is corresponding method without/with the loop elimination, respectively. The EDA-RL with the linear-chain model and the loop elimination could not solve these maze problems well, where resultant fitness at the last generation is around 800. The reason of this is that the maze problems have several perceptual aliasing states. As a consequence of the structure search, the probabilistic model is partially changed from the independent features to the 3-clique model if need arises. Therefore, the performance of the proposed method cannot outperform the 3-clique model in principle. However, in the case of the larger maze problem and that the number of episodes for policy estimation is 512, the 3-clique model could not solve the larger maze at all while the proposed method with the loop elimination show better performance. This is caused by the number of features summarized in Table 14.1. That is, due to the structure search, where it adds new feature only for perceptual aliasing states, the proposed method seems to work well. In the case of the 3-clique model, on the other hand, the agents have to take account the previous states even if the agents are in non-perceptual aliasing states.

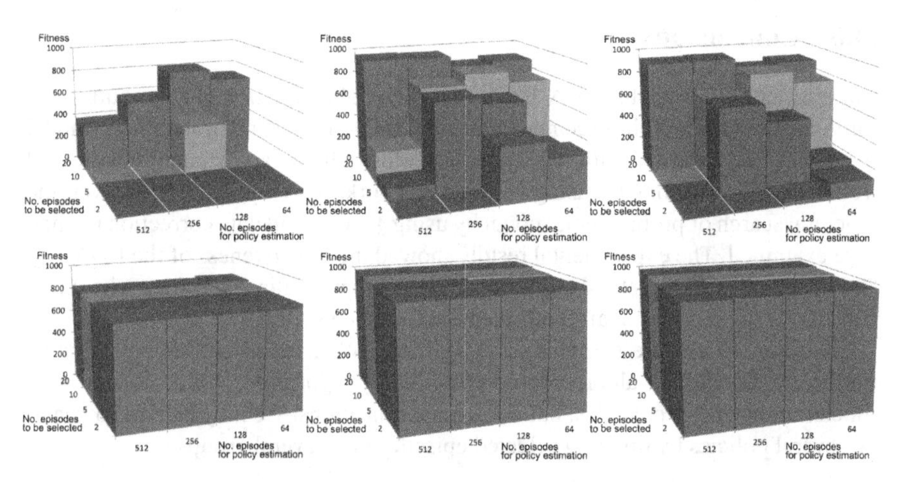

Fig. 14.8 Experimental results for the smaller maze problem: (UPPER) without loop elimination; (LOWER) with loop elimination: (LEFT) Linear Chain Model; (CENTER) 3-clique Model; (RIGHT) Proposed Method (Structural Search)

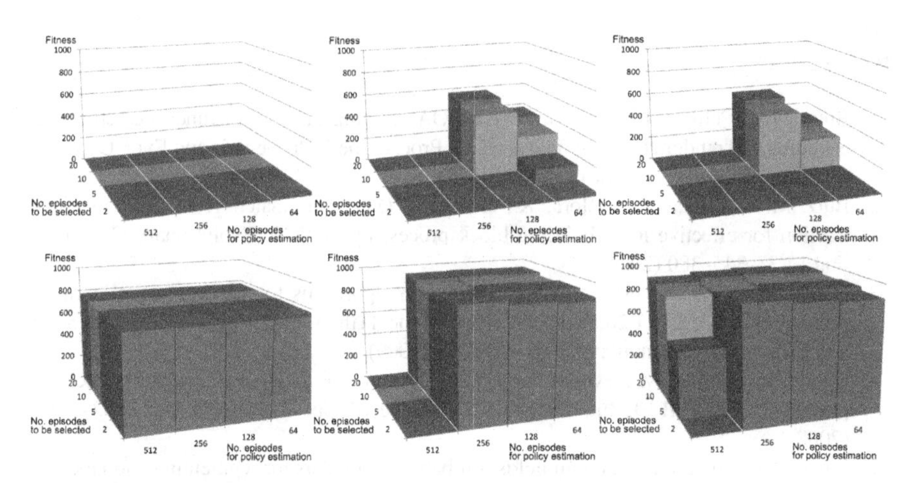

Fig. 14.9 Experimental results for the larger maze problem: (UPPER) without loop elimination; (LOWER) with loop elimination: (LEFT) Linear Chain Model; (CENTER) 3-clique Model; (RIGHT) Proposed Method (Structural Search)

Table 14.1 The number of features in the probabilistic models

Model	Smaller Maze	Larger Maze
Linear-Chain	41	47
Ind.-Features	25	31
3-clique	217	255
Struct. Search	59.5	82.5
Struct. Search w loop elim.	72.9	98.3

14.6 Conclusions

This chapter introduced the EDA-RL: an extension of Estimation of Distribution Algorithms for solving Reinforcement Learning Problems. As the probabilistic model of the EDA-RL, the Conditional Random Fields, which can estimate conditional probability distributions by using Markov Network, are employed. Moreover, the structural search of probabilistic model by using χ^2-test, and data correction method were examined. The experimental results showed the effectiveness of the EDA-RL. The structural search method can resolve the perceptual aliasing states, where it is difficult for the conventional Reinforcement Learning methods.

The future works are summarized as follows: We need to compare with some reinforcement learning algorithms or evolutionary algorithms, which change their policies at the end of episodes, but not at state transitions. We may be able to generate several policies by dividing selected episodes into several groups.

Acknowledgement. This work was partially supported by the Grant-in-Aid for Exploratory Research and the Grant-in-Aid for Young Scientists (B) of MEXT, Japan (21700254 and 23700267).

References

1. Butz, M.V., Pelikan, M.: Studying XCS/BOA learning in boolean functions: structure encoding and random boolean functions. In: Proc. of the 2006 Genetic and Evol. Comput. Conf., pp. 1449–1456 (2006)
2. Butz, M.V., Pelikan, M., Llorá, X., Goldberg, D.E.: Automated global structure extraction for effective local building block processing in XCS. Evolutionary Computation 14(3), 345–380 (2006)
3. Handa, H.: EDA-RL: estimation of distribution algorithms for reinforcement learning problems. In: GECCO 2009: Proceedings of the 11th Annual Conference on Genetic and Evolutionary Computation, pp. 405–412 (2009)
4. Handa, H., Isozaki, M.: Evolutionary fuzzy systems for generating better Ms.PacMan players. In: 2008 IEEE International Conference on Fuzzy Systems, pp. 2182–2185 (2008)
5. Lafferty, J.: Conditional random fields: Probabilistic models for segmenting and labeling sequence data. In: Proceedings of 18th International Conference on Machine Learning, pp. 282–289. Morgan Kaufmann (2001)
6. Miyazaki, K., Yamamura, M., Kobayashi, S.: On the rationality of profit sharing in reinforcement learning. In: Proc. 3rd International Conference on Fuzzy Logic, Neural Nets and Soft Computing (IIZUKA 1994), pp. 285–288 (1994)
7. Ono, I., Nijo, T., Ono, N.: A genetic algorithm for automatically designing modular reinforcement learning agents. In: Proc. of the Genetic and Evol. Comput. Conf., pp. 203–210 (2000)
8. Sutton, C., Mccallum, A.: An introduction to conditional random fields for relational learning. In: Getoor, L., Taskar, B. (eds.) Introduction to Statistical Relational Learning, ch. 4, pp. 93–128. MIT Press, Cambridge (2007)

9. Sutton, R.S., Barto, A.G.: Reinforcement Learning: An Introduction. MIT Press (1998)
10. Watkins, C., Dayan, P.: Technical note: Q-learning. Machine Learning 08, 279–292 (1992)
11. Yamazaki, A., Shibuya, T., Hamagami, T.: Complex-valued reinforcement learning with hierarchical architecture. In: Proc. of IEEE International Conference on Systems, Man, and Cybernetics, pp. 1925–1931 (2010)

Index